T0122943

NEURAL COMPUTATION IN HOPFIELD NETWORKS AND BOLTZMANN MACHINES

Neural Computation in Hopfield Networks and Boltzmann Machines

James P. Coughlin
Robert H. Baran

Newark: University of Delaware Press
London and Toronto: Associated University Presses

Associated University Presses
440 Forsgate Drive
Cranbury, NJ 08512

Associated University Presses
25 Sicilian Avenue
London WC1A 2QH, England

Associated University Presses
P. O. Box 338, Port Credit
Mississauga, Ontario
Canada L5G 4L8

The paper used in this publication meets the requirements of the American National Standard for Permanence of Paper for Printed Library Materials Z39.48-1984.

Library of Congress Cataloging-in-Publication Data

Coughlin, James P.
Neural computation in Hopfield networks and Boltzmann machines/ James P. Coughlin, Robert H. Baran.
p. cm.
Includes index.
ISBM 0-87413-464-1 (alk. paper)
1. Neural networks (Computer science) 2. Machine theory.
I. Baran, Robert H. II. Title.
QA76.87.C69 1994
006.3–dc20 94-23048
CIP

Printed in the United States of America

CONTENTS

PREFACE

When asked to define "intelligence," Marvin Minsky once offered some introspective observations along the following lines: Intelligent behavior is anything that I would like to be able to do, but can't (due to my own limitations). By studying and practicing such behavior, I gradually learn to understand and copy it. But the more capable I become, the easier — and less "intelligent" — it seems.

No doubt this fact of human nature afflicts the present text in the forms of carelessness, unevenness, irreverence, and unfortunate omissions. How could we have maintained the exuberant fascination that inspired the study leading to this result?

It was hoped that the present volume might have been as prescient and concise as the papers by Hopfield that established the existence of the subject. It would have been nice to share the intuitive vision and originality of Hinton and Sejnowski, whose development of the "Boltzmann machine" from the Hopfield model is the main concern of the second half. Add to these desiderata the unencumbered authority of Feller's *Introduction to Probability Theory and Its Applications* and the pedagogical good humor and effectiveness of Drake's *Fundamentals of Probability Theory* — and the reader will see how far short of our ideals we have fallen.

Our initial intention was to write an account of neural networks suitable for graduates and advanced undergraduates. We began with the Hopfield model and had planned to survey a broad range of paradigms. Each time we became comfortable with one of the mathematical principles of collective computation, certain questions arose; and ultimately we devoted most of three years to symmetrically interconnected networks of binary neurons.

The practical utility of this book as the basis of a popular course probably fell by the wayside when we persevered in the study of the stochastic model that Hopfield abandoned in 1984. While most students in science, engineering, and mathematics must study probability, most go on to practice strict determinism. Yet this perseverance was inspired by an appreciation for the remarkable theoretical consequences of the model that Hinton and Sejnowski had already foreseen in 1983.

The challenge of teaching the most far-reaching consequences of stochastic models without a thorough study of their mathematical fundamentals is completely appreciated. If this text has no other redeeming virtues, perhaps it can serve as a preview of statistical mechanics by way of neural networks — and remind the sophisticated reader that, sooner or later, it is the simplest notions that propel the most profound advances in science and technology. To illustrate this point, one needs only to ponder the consequences of "the atomic theory" — and tremble at the prospect that, someday, someone will split the neuron!

ACKNOWLEDGMENTS

No book is possible without the collaboration of many people and this one is no exception. The authors would like to thank first and foremost our wives, Arlene Coughlin and Virginia Baran, who bore the position of "book widow" with a great deal of grace. We would also like to thank Ms. Carolyn Westbrook for assistance with the artwork as well as Ms. Dolly Myers and Ms. Joyce Dorfler for typing.

NEURAL COMPUTATION IN HOPFIELD NETWORKS AND BOLTZMANN MACHINES

1. Elements of the Neural Network Paradigm

Neural networks, like Bradbury's Martians, are different things to different people. Here we consider artificial neural networks that are based on connectionist models. The essential building blocks of the connectionist model are:

(1) A set of neuronlike units, which may be called neurons, units, or (Ising) spins. These units can always be indexed serially by i = 1, 2, ...,N. Then the state of the network at an unspecified time is $\mathbf{x} = (x_1,\ldots,x_N)$. Sometimes it will be convenient to adopt a different indexing scheme (as in the traveling salesman problem).

(2) A matrix of synaptic weights. These neurons are connected to one another by connections called synapses. The strengths of these connections can be listed in a matrix **W**. When the neurons are labeled serially from 1 to N, W_{ij} is the strength of the synaptic connection *from* j *to* unit i. When the weight is positive, the connection is excitatory, and when negative it is inhibitory.

(3) Some (neuro)dynamic assumptions. The dynamics must describe the evolution of the state of the system in time.

The Connectionist Neuron

The simplest neurodynamic system is described by the connectionist neuron shown in fig. 1.1. The total "activation energy" coming into neuron #i is obtained by adding together the strengths of the synapses connecting it to all other active neurons. If the synaptic weight is negative, the sum is decreased. If a neuron is totally inactive, its state is $x_i = 0$ (versus $x_i = 1$ when it is fully on). The presynaptic activity constituted by the other units is formed into a weighted sum:

$$y_i = \sum_{j=1}^{N} W_{ij}\, x_j \qquad \textbf{(1)}$$

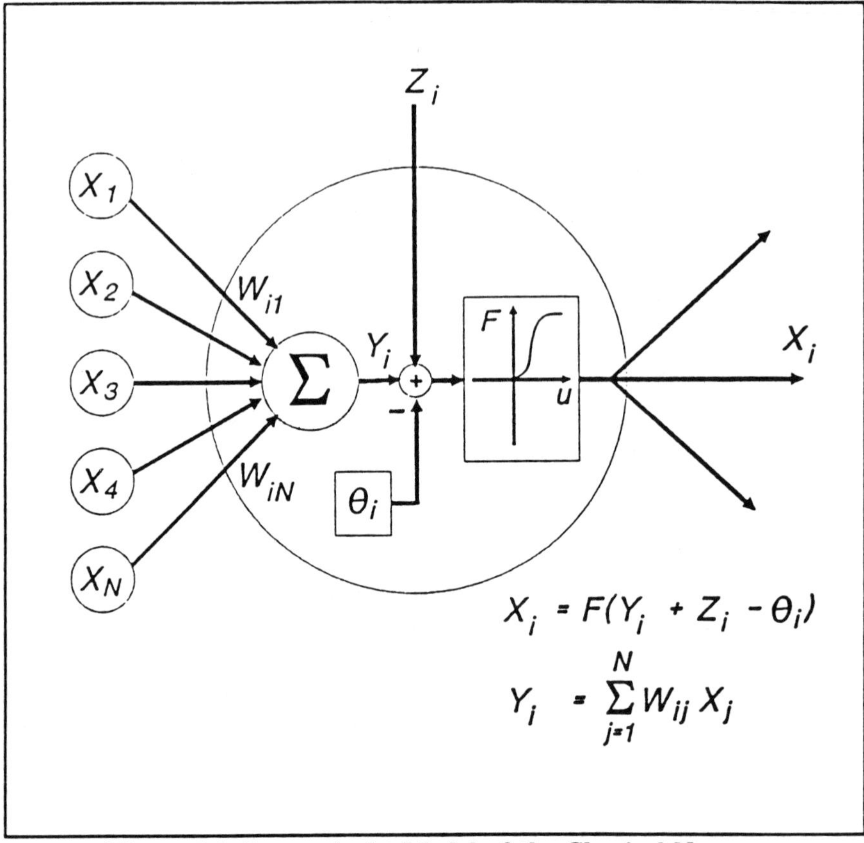

Figure 1.1 Connectionist Model of the Classical Neuron

There may be an input to the ith neuron from sources other than the neurons thus considered. Let this externally applied input or bias signal be denoted z_i. Then the total input to unit i is $u_i = y_i + z_i$. The neuron compares the total input to a threshold θ_i by forming the difference $u_i - \theta_i$; and its output — its state of activation (or "state") is:

$$x_i = F_i\ (u_i - \theta_i)\quad , \tag{2}$$

where $F_i(.)$ is a bounded (increasing), real-valued function that generates the mean firing rate, x_i, i.e., the average rate of action potential generation for the given input and threshold. Whenever F(.) appears without a subscript, the neurons of the network are assumed identical in this respect.

Connectionist models use either graded neurons or binary threshold units (fig. 1.2). The latter are also called *McCulloch-Pitts neurons*. Following

Hopfield (1982) we use the term McCulloch-Pitts neuron specifically when F is the unit step (or Heaviside) function:

$$\mathcal{H}(u) = \begin{cases} 1 & \text{if } u > 0 \\ 0 & \text{if } u \leq 0. \end{cases}$$

Neurons that follow this law are either ON or OFF.

We use the term "sign neuron" when F is the sign function:

$$sgn(u) = \begin{cases} +1 & \text{if } u > 0 \\ -1 & \text{if } u \leq 0. \end{cases}$$

Note the sign/step transformations: If $v = \mathcal{H}(u)$ and $s = sgn(u)$, then $v(t) = (s(t) + 1)/2$ and $s(\mathbf{x}) = 2v(\mathbf{x}) - 1$. These binary threshold units dominated the work of early pioneers like Caianiello (1961, 1986) and Rosenblatt (1960, 1961, 1962). Hopfield (1982) also used McCulloch-Pitts neurons, which likewise figure prominently in subsequent chapters. Graded neurons have received considerable attention in the 1980s. The (logistic) sigmoidal activation function:

$$F(u) = \frac{1}{1+e^{-bu}} \qquad \textbf{(3)}$$

(fig. 1.2 ii) was used by Rumelhart et al. (1987) in presenting the derivation of their "backpropagation" technique. Note that the state of activation given by this sigmoid lies in the unit interval ([0,1]) and that, in the limit as $b \to \infty$, it approaches the unit step.

In the physics literature, where the sign neuron is an Ising spin (as discussed in chapter 4), the form $F(u) = \tanh(bu)$ is commonly used.

<u>Problem</u> 1.1: Let $v = F(u)$ where F is as given by equation (3). Invoke the step-to-sign transformation and express the result in closed form.

Solution:

$$\begin{aligned} s &= 2v-1 = 2F(u) - 1 \\ &= (1 - e^{-bu})/(1 + e^{-bu}) \\ &= \tanh(bu/2). \end{aligned}$$

The sign neuron is the limiting case of this (as $b \to \infty$).

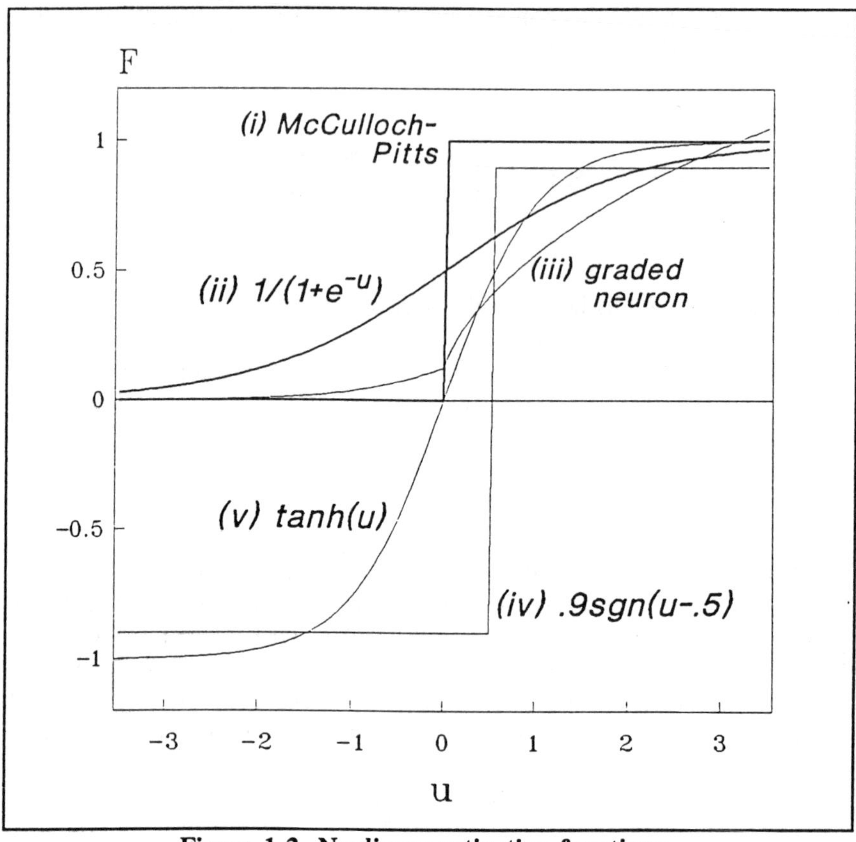

Figure 1.2 Nonlinear activation functions.

Simple Neural Networks

Connectionist neurons are usually represented in network diagrams by circles or occasionally by squares. A line drawn between them represents a synapse whose direction is indicated by an arrowhead. It is nice (but not necessary) to draw inhibitory connections with dashed or dotted lines. The three-unit network in fig. 1.3 (left) has a bias signal of $+\epsilon$ applied to the top-most unit, and — following Rumelhart et al. (1987) — a threshold of $+\delta$ in the right-most unit. Here we shall use a different convention, illustrated on the right of fig. 1.3, where the indexed states of the units appear in the circles. This leaves no place for the thresholds; but, since a threshold of $+\delta$ is always equivalent to an externally applied bias of $-\delta$, and vice versa, we can, without loss of generality, dispense with thresholds by representing them as biases.

The network in fig. 1.3 features a reciprocal synapse, i.e., an instance of $W_{ij} = W_{ji}$. Rather than drawing two directed links, each of strength b, between units 2 and 3 (as on the left), we draw a single link of strength b, either with arrowheads on both ends or with no arrowheads at all (as on the right). When the subject is collective computation (as in Hopfield nets or Boltzmann machines), the synapses are always reciprocal and in subsequent chapters we restrict ourselves to networks with reciprocal synapses and their corresponding symmetric weight matrices.

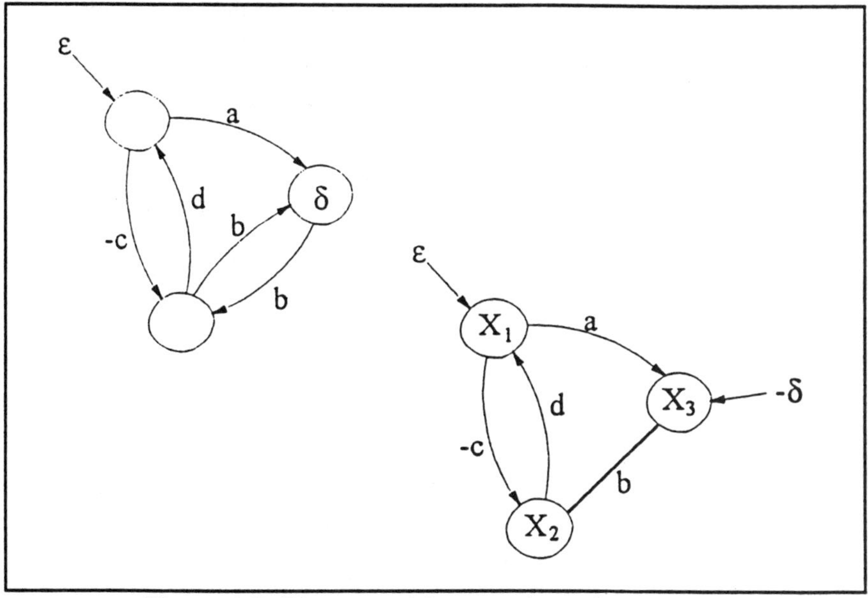

Figure 1.3 Neurons with nonreciprocal synapses.

The network diagram is a somewhat elaborate way to present the information it contains. Fig. 1.3, for instance, can be represented by the matrices:

$$W = \begin{bmatrix} 0 & d & 0 \\ -c & 0 & b \\ a & b & 0 \end{bmatrix} \quad \text{and} \quad z = \begin{bmatrix} \epsilon \\ 0 \\ -\delta \end{bmatrix} .$$

The purpose of the network diagram is to clarify the logical behavior that results from given weights and inputs. For instance, what would be the effect of clamping certain units [i.e., fixing their firing rates independently of the other neurons] upon the activity of the unclamped units? A McCulloch-Pitts neuron, for example, is clamped ON when the bias signal is sufficiently excitatory to assure that the state is $x = 1$ (ON) irrespective of the states of the other units; and the unit is clamped OFF when the bias strongly inhibits the unit. It will not be necessary to indicate the bias explicitly when the clamp is understood.

Example: April, the assistant secretary for snap judgments, advises the secretary to say "yes" or "no" to questions of policy that arise from time to time. The assistant secretary has two under assistants, Clark and Les, who give her inputs of the same nature. After several months on the job, she discovers that Clark always gives correct advice, and Les is almost always wrong; but Les will occasionally affirm Clark's "yes" to try to confuse her. The judgments of these four individuals can be represented by four McCulloch-Pitts neurons, $x = 1$ corresponding to "yes" and $x = 0$ to "no". The sketch shows a weight assignment that solves the problem faced by the secretary. For instance, if Clark says "yes" and Les says "no," units #1 and #2 are clamped ON and OFF, respectively; and the net input to April (#3) is $1+\epsilon+\delta$,. Obviously the same result could be obtained by transferring Les to some kind of bureaucratic Siberia and following Clark alone. The advantage of having Les around is that when,

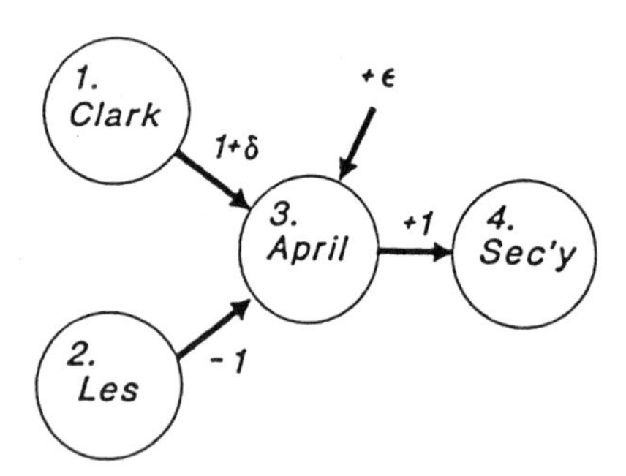

in Clark's absence, he says "yes," April can always override her natural inclination as a yes-person, indicated by the bias of $+\epsilon$, and go the other way.

Problem 1.2 (The Exclusive-Or): April is promoted to secretary and she hires an assistant, June, who is unquestionably loyal. But June replaces Clark and Les with her own advisers, Bill and Don, who seldom agree on anything. In fact, when either Bill or Don says "yes," this is the correct advice. If both men agree, however, something must be wrong with the proposition, and the correct judgment is "no". (The perceptive reader will observe that this is a simple logical circuit designed to implement the exclusive "or.")

Solution: As Minsky and Papert pointed out, there is no pair of weights (W_{31} and W_{32}) that will cause June (#3) to give correct advice to April (#4). (The reader may be as surprised as we were to discover this.) But if April convenes a meeting of all three subordinates, and takes all their responses into account, the weight assignment shown below suffices. June turns ON only when either Bill or Don says "yes"; and April copies June except when both men are affirmative. The table shows y_3 and y_4, the net inputs to the units representing June and April, as functions of the (clamped) states of the units for Bill and Don.

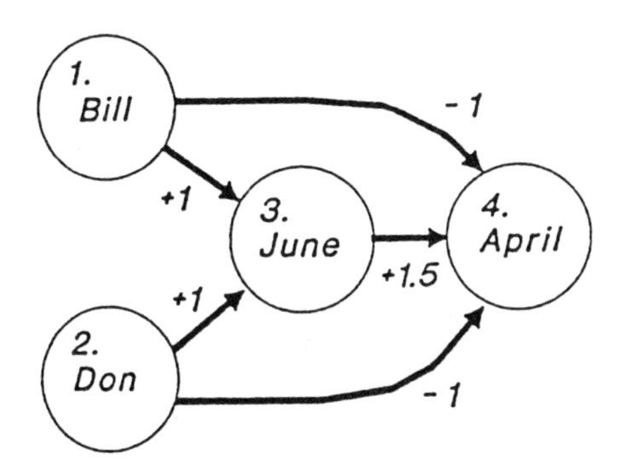

		---- Inputs ----		
Bill	Don	June	April	Output
x_1	x_2	y_3	y_4	x_4
0	0	0	0	0
0	1	+1	+0.5	1
1	0	+1	+0.5	1
1	1	+2	-0.5	0

Moreover, if June adopts a skeptical disposition, signified by the bias of -1.5 in the lower left quadrant of fig. 1.4, then the network drawn by Rumelhart et al. (1987, 321) is also effective.

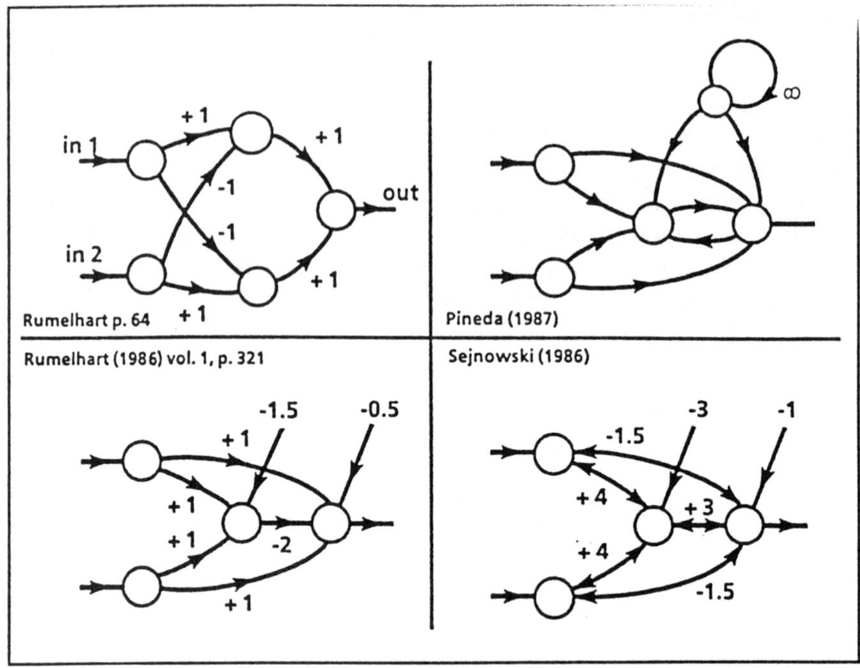

Figure 1.4 Four PDP Realizations of the XOR.

The Perceptron and Backpropagation

Historically speaking, the first workable artificial neural net was probably the perceptron of Frank Rosenblatt. Rosenblatt (1961/62, 100) showed the existence of universal perceptrons, capable of correctly classifying every element of a "stimulus world." Consider the "elementary perceptron" in which there is a single response (R) unit, a binary threshold unit that divides the stimulus world into two classes. Some sensory (S) units, which are McCulloch-Pitts neurons, represent the stimuli. Between the S-layer and the R-unit there will be some association (A) units, also of the McCulloch-Pitts type. In the current parlance, these A-units comprise the "hidden layer." Fig. 1.5 is a facsimile of Rosenblatt's sketch depicting the elementary S-A-R perceptron.

Let there be one hidden unit for every possible pattern (or input vector). Let the ith hidden unit have an excitatory connection (+1) from every input unit that is ON in the ith pattern and an inhibitory (-1) connection from each input unit that is OFF. Also assign a threshold to each hidden unit, equal to one less

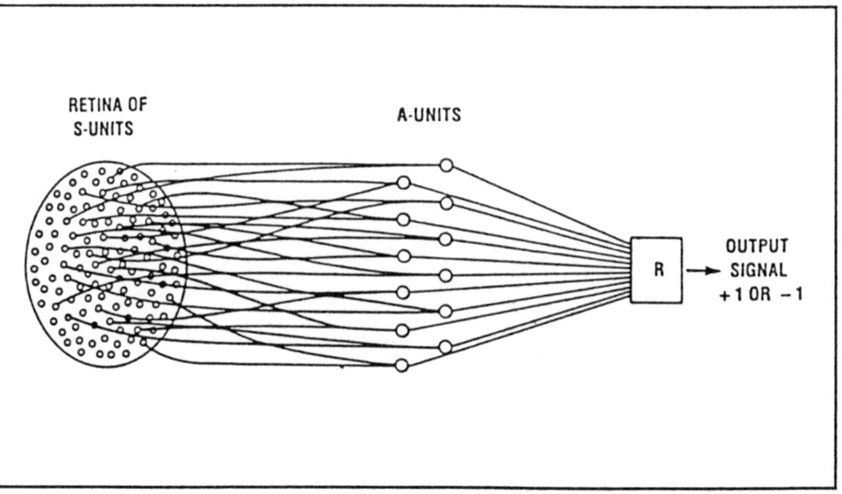

Figure 1.5 Network organization of a typical elementary perceptron.

han the number of excitatory connections it receives from input units. Then the th hidden unit will always turn ON when the input layer is clamped to the ith attern, remaining OFF for all other inputs.

Problem 1.3: Diagram an elementary perceptron for solving the xclusive-or (XOR) problem. (The reader may wish to review problem 1.2 as vell as the definition of perceptron.)

Solution: It is not really necessary to include a hidden unit that turns)N to the all-zero pattern. Therefore let there be three hidden units with 0-1 inary states x_1 through x_3. Fig. 1.6 shows the universal perceptron solution to he XOR problem using externally applied biases instead of thresholds.

This solution can be further simplified. Elimination of hidden unit #2 loes not change the truth table for this neural logic gate. (The net inputs y_1 and $_3$ are identically zero when both inputs are clamped ON.) Thus we reduce the etwork to five units with exactly the same weights as shown in the upper left uadrant of fig. 1.4.

As Rosenblatt noted, the existence of such universal perceptrons is quite nexciting in view of the proliferation of hidden units in any problem large nough to be of practical interest. Most of Rosenblatt's three-layer, series-oupled (S-A-R) perceptrons initially derived the weights of their input layer-to-idden layer unit (S-A) connections from a random number generator by way

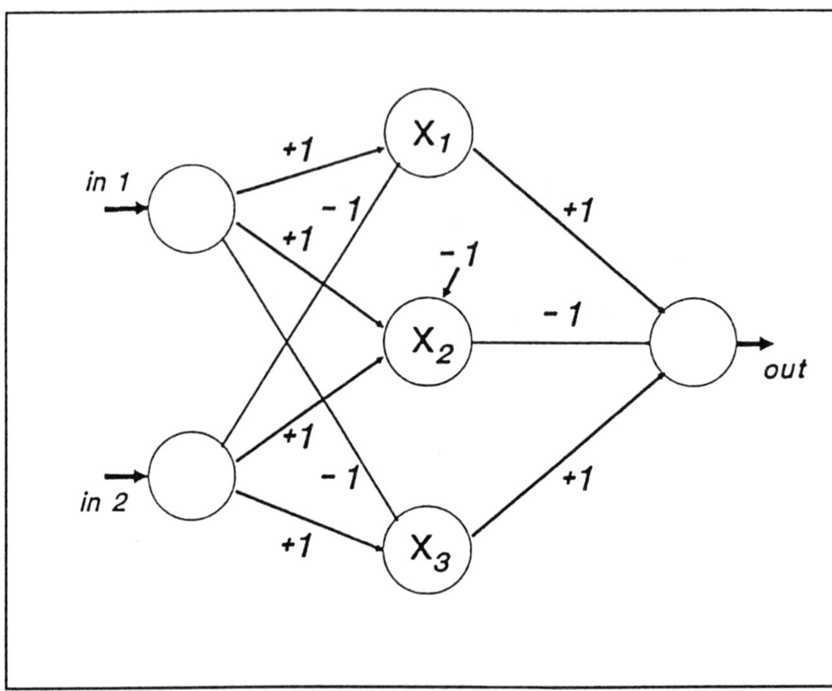

Figure 1.6 Perceptron with the input layer on the left, output layer on the right.

of stochastic rules. The hidden layer-to-output layer unit (A-R) connections, initially with random weights, were modified incrementally in the course of many passes through a training set (consisting of prescribed i/o pairs) using a reinforcement control procedure (or learning algorithm). The asymptote of the individual perceptron learning curve was obviously limited by the suitability of the S-A weights.

Rosenblatt (1961/62, p. 292) also used a "*back-propagating error correction procedure*" for training the S-A weights. The essence of this backpropagation technique was a brief list of rules for *assigning errors to hidden units* based on their interactions with output units that assumed the wrong state in response to the training input. Then the weight modification formula for the connections terminating on hidden units was formally the same for those terminating on output units. Only the error computation was different. Rosenblatt's backward error propagation rules were probably doomed because they were of the stochastic (i.e., probabilistic) type. Perhaps (paraphrasing Einstein) neural networks should not play dice (at least while learning).

The revival of the perceptron architecture in the 1980s was made possible by a different backpropagation algorithm, based on the (tacit) assumption of graded neurons (see p. 2) and the mathematics of gradient descent in squared error. Hecht-Nielsen attributes the discovery of this deterministic backpropagation to Werbos; but Rumelhart et al. (1987) brought it to the world's attention, and Sejnowski et al. (1986) made it popular with powerful demonstrations, especially NETtalk, the system that learned to read aloud in English. Rosenblatt had offered a proof to the effect that any strictly deterministic training procedure could not be guaranteed to converge to the correct weights — despite the existence of such a solution for the given numbers of units in each layer. In modern terms, the weight matrix can attain a local minimum (of the error function) that strict gradient descent cannot escape. Yet the new backpropagation techniques have been highly effective in almost every published account; and the primary vendors of backpropagation accelerators today dismiss the objection.

Training the Neural Network

Between the death of Frank Rosenblatt and the first (child-like) utterances of NETtalk, prominent investigators redefined the intent and significance of the neural network paradigm in terms of *Parallel Models of Associative Memory*. A collection of papers under this title, edited by Hinton and Anderson (1981) provides a view to the state of the art just prior to Hopfield's first celebrated contribution. Grossberg's (1988) account of developments during this period is must reading for every serious student; but it goes perhaps too far in emphasizing the extent to which Hopfield's discoveries were foreshadowed by prior work.

Kohonen (1977, 1981) had made it clear how Hebbian learning captured statistical correlations among suitably represented data items in the form of a weight matrix:

$$W_{ji} = (1/M)\sum_{m=1}^{M} a_i^{(m)} b_j^{(m)} \qquad (4)$$

where $a_i^{(m)}$ is the actual activation of the ith input unit when the mth stimulus is impressed on the net, and $b_j^{(m)}$ is the actual activation of the jth output unit when the correct response is likewise impressed. The weights are modified by:

$$W_{ji} = W_{ji}^{o} + \Delta_m W_{ji}$$

where:

$$\Delta_m W_{ji} = \eta a_i^{(m)} b_j^{(m)}$$

and η is positive whenever the units' activities agree in sign and negative when they disagree. If each of the M stimulus/response (or i/o) pairs is presented once, the end result is the same as equation (4) after setting $\eta = 1/M$.

$$W_{ji} = \sum_{m=1}^{M} \Delta_m W_{ji} = \sum_{m=1}^{M} \eta a_i^{(m)} b_j^{(m)}$$

Now when a vector **z** of bias signals is applied to the input units, they assume states **x**=F(**z**); and the response of an output unit of the perceptron is:

$$x_j = F\,[\,\sum_i W_{ji} F(z_i)\,] \quad .$$

If the units are sign neurons (see page 3) then:

$$x_j = sgn[\sum_m b_j^m (1/M) \sum_i a_i^m sgn(z_i)]$$

after substituting equation (4) for the weights and reversing the order of the summations. If the input vector is (any positive multiple of) one of the training patterns — say the qth such pattern — then:

$$\sum_i a_i^m sgn(z_i) = \sum_i a_i^m a_i^q \quad .$$

The vectors {$\mathbf{a}^m$: m=1,...,M} are said to be <u>orthonormal</u> if:

$$(1/M) \sum_i a_i^m a_i^q = \delta_{mq} = \begin{cases} 1 & if\ m=q \\ 0 & if\ not \end{cases}$$

(Kronecker's delta function). Thus the orthogonality of the input vectors results in outputs of the form:

$$x_j = sgn(\sum_m b_j^m \delta_{mq}) = sgn(b_j^q) = b_j^q \; .$$

In this way the storage prescription (4) associates the correct output vector with the input. (Kohonen's 1977 work actually assumed linear neurons and led to some profound results based on the algebra of pseudo-inverses.)

Problem 1.4 : Design a weight matrix that associates the three (transposed) input patterns (+ - + -), (- - + +), and (+ + + +) with the output patterns (+ - -), (- + -), and (- - +).

Solution: See fig. 1.7. From equation (4):

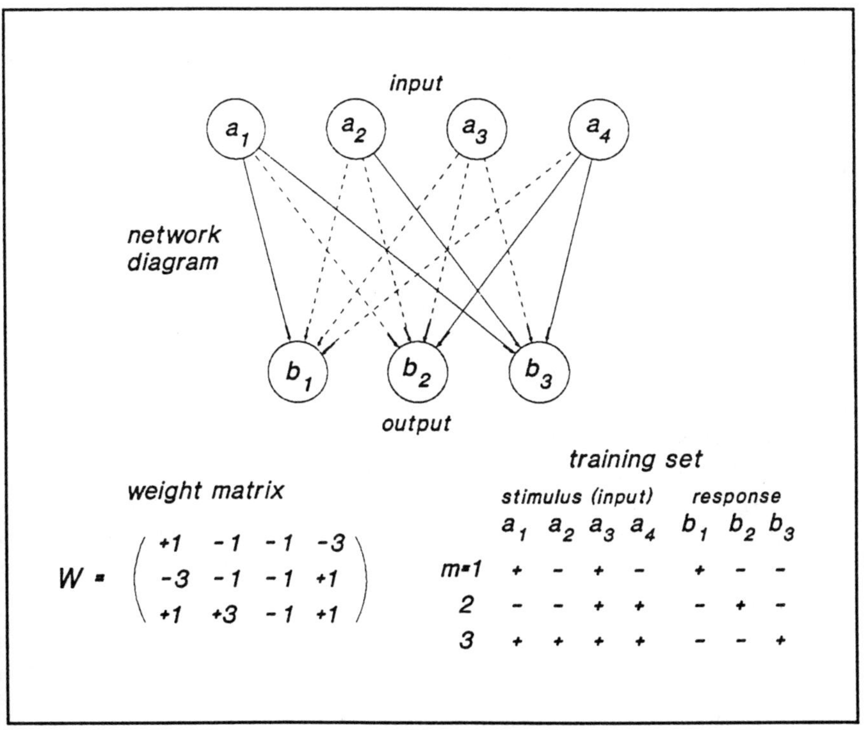

Figure 1.7 Associative memory made with sign units and a weight matrix (W) derived from a training set.

$$3W_{ji} = a_i^{(1)}b_j^{(1)} + a_i^{(2)}b_j^{(2)} + a_i^{(3)}b_j^{(3)} .$$

and 3**W** is as shown in the figure. Instead of applying **W** to each **a** separately we can multiply **W** by the matrix whose columns are the **a**'s. Applying the

$$W(a^1, a^2, a^3) = \begin{pmatrix} +1 & -1 & -1 & -3 \\ -3 & -1 & -1 & +1 \\ +1 & +3 & -1 & +1 \end{pmatrix} \begin{pmatrix} +1 & -1 & +1 \\ -1 & -1 & +1 \\ +1 & +1 & +1 \\ -1 & +1 & +1 \end{pmatrix}$$

$$= \begin{pmatrix} +4 & -4 & -4 \\ -2 & +5 & -4 \\ -4 & -4 & +4 \end{pmatrix}.$$

signum function to each element of the result gives a matrix whose columns are the elements of **b**.

In an **auto-associative memory** the input and output units are in one-to-one correspondence. The weights might be given by equation (4) with **b** = **a**. Fig. 1.8, for example, uses the same three input patterns as in problem 1.4.

The Hopfield Net

Hopfield (1982) introduced a new kind of content-addressable memory (CAM) in which the input and output are represented by the same collection of neurons. Patterns are stored in the "Hopfield net" by the same outer product rule, equation (4), specialized to the auto-associative case. In particular, let $(x_1^m,\ldots,x_N^m) = \mathbf{x}^m$ be a vector of binary 0-1 components that the network is to remember. The pattern set $\{\mathbf{x}^m, m=1,\ldots,M\}$ contains M such N-bit vectors. With the step-to-sign transformation $a_i^m = 2x_i^m-1$, the outer product rule is written as:

$$W_{ij} = (1/M)\sum_{m=1}^{M} (2x_i^m-1)(2x_j^m-1) \qquad \textbf{(5a)}$$

if $i \neq j$, but 0 if $i = j$:

$$W_{ii} = 0. \qquad \textbf{(5b)}$$

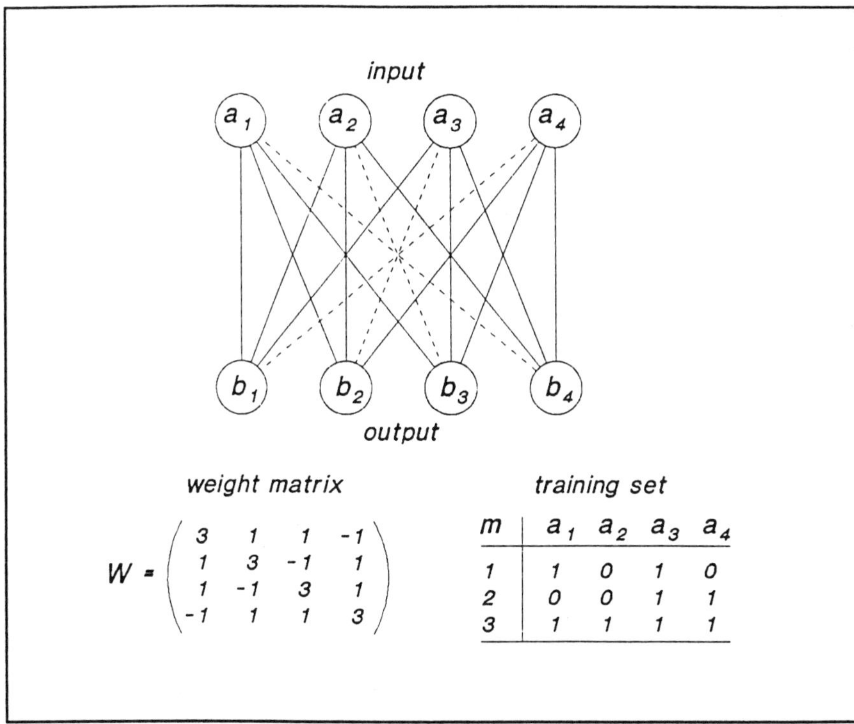

Figure 1.8 An auto-associative memory with indicated weight matrix (W) by application of the outer product rule to the training set (below, right).

The purpose of this storage prescription is to create a network, the stable states of which coincide with the desired patterns. Pineda (1987) has remarked that two kinds of weight matrices guarantee the existence of stable states in a neural network. The first kind is the lower-triangular block matrix, which gives rise to a layered, feed-forward architecture. Most of the examples shown thus far — particularly the perceptron and the two-layer associative memories — have had this structure. (The **W** of the associative memory takes on a lower-triangular form if the units are indexed serially, e.g., from 1 to 7 in fig. 1.7). The existence of stable states in these networks was so obvious as to require no explicit discussion of the dynamical process by which an initial state evolves into a (final) stable state. One merely propagates the activation, starting with the clamps on the input units, forward to the output via the hidden layers (if any).

The other kind of weight matrix that guarantees "global pattern formation," (understood as the convergence to stable states), is the symmetric form, $W_{ij}=W_{ji}$, with $W_{ii}=0$. Application of equation (5) produces the reciprocal connections that correspond to weight matrix symmetry. The stable states of the Hopfield net will satisfy the fixed point equations:

$$\boldsymbol{x} = \mathcal{H}\,(\boldsymbol{Wx} + \boldsymbol{z}) \tag{6}$$

for McCulloch-Pitts neurons. In the absence of any biases, this is:[1]

$$x_i = \mathcal{H}\left\{(1/M)\sum_{m=1}^{M}(2x_i^m - 1)[\sum_{j=1}^{N}(2x_j^m - 1)x_j]\right\}$$

after substituting equation (5a) and reversing the order of summation. The reader may derive this as an exercise. Let N be an even number and suppose that every pattern has the same activity, A:

$$A(m) = \sum_{i=1}^{N} x_i^m = N/2 \quad . \tag{7}$$

Then orthonormal pattern sets will have:

$$\sum_j (2x_j^m - 1)x_j^q = \sum_j 2x_j^m x_j^q - N/2 = N(\delta_{mq} - 1/2) \quad . \tag{8}$$

Hence:

$$x_i = \mathcal{H}\,[(N/M)\sum_m (2x_i^m - 1)(\delta_{mq} - 1/2)] = x_i^q$$

and the stored patterns are stable. Moreover, if the patterns are random, each produced by N independent Bernoulli trials with success probability 1/2 — what non-statisticians call "coin tosses" — then the pseudo-orthogonality of the patterns will make the right hand side of equation (8) equal to the average value of the left.

Hopfield noted (fig. 1.9) that "any physical system whose dynamics...is dominated by stable states to which it is attracted can be regarded as a general CAM. Such a physical system will be potentially useful if any prescribed set of states can be made the stable states of the system." Clearly the outer product rule has some value for satisfying the requirement of utility. It remained to show that the network is "attracted" to these stable states. Hopfield's proof

---

[1] Refer to p.3 for the definition of $\mathcal{H}$(u).

The Model

"Each neuron i has two states like those of McCulloch and Pitts:" $V_i = 0$ (resting) and $V_i = 1$ (firing). T_{ij} is the strength of the connection to neuron i from neuron j. "Each neuron readjusts its state randomly in time but with mean attempt rate W, setting

$$\begin{matrix} V_i = 1 \\ V_i = 0 \end{matrix} \quad \text{if } \sum_j T_{ij} V_j \quad \begin{matrix} > U_i \\ < U_i \end{matrix} \text{,"}$$

where U_i is a fixed threshold.

Storage Prescription

"Suppose we wish to store a set of states V^s, s = 1,2,...,n. Use

$$T_{ij} = \sum_s (2V^s_i - 1)(2V^s_j - 1)$$

but with $T_{ij} = 0$."

Collective Behavior

"The model has stable limit points. Consider the special case $T_{ij} = T_{ji}$ and define

$$E = -1/2 \sum_{i \neq j}\sum T_{ij} V_i V_j.$$

ΔE due to ΔV is given by

$$\Delta E = -\Delta V \sum_{j \neq i} T_{ij} V_j.$$

Thus the model causes E to be a monotonically decreasing function. State changes will continue until a least (local) E is reached."

Figure 1.9 The main points stated by Hopfield (1982).

consisted of observing that each network state vector has a "computational energy" that can only decrease through the *asynchronous* operation of the neurons. Since the energy (being bounded) cannot decrease indefinitely, the network ultimately settles into stable states at (local) energy minima. The storage prescription is a means of creating "energy valleys" about the desired coordinates in state space. More will be said about this in chapter 3.

Problem 1.5: Use the outer product rule to create a CAM that stores the three four-bit patterns labeled {**a**} in figs. 1.7 and 1.8. Verify the stability of the patterns. Are there any other stable states?

Solution: Although only the upper triangle of the symmetric weight matrix needs to be displayed, we give the entire matrix for clarity's sake. The network diagram is shown in fig. 1.10.

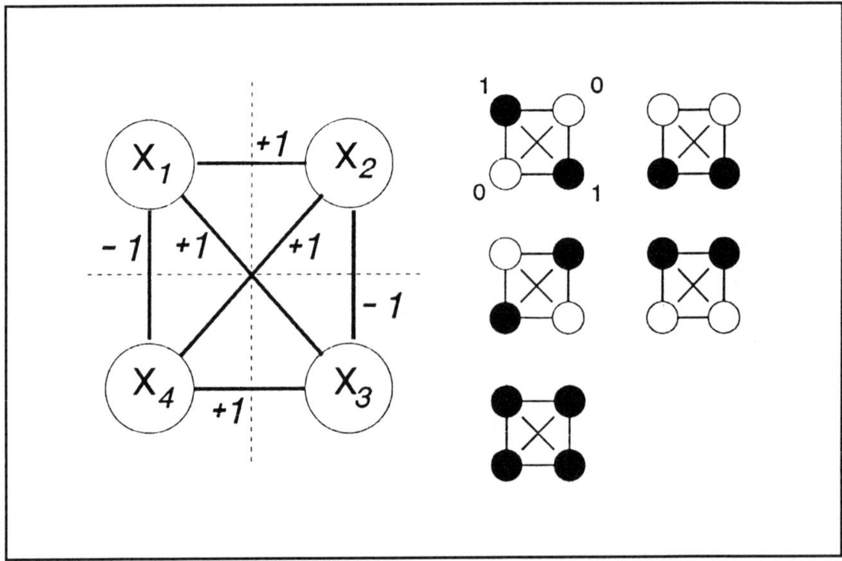

Figure 1.10 Network diagram (left) of a C.A.M. with four McCulloch-Pitts neurons.

$$W = \begin{pmatrix} 1 & 1 & -1 \\ 1 & -1 & 1 \\ -1 & 1 & 1 \end{pmatrix}.$$

The stable states are indicated in the figure with the black units being ON. As can be seen in fig. 1.10 the stable states (right) include the three stored patterns (top row) and two undesired states, the stability of which results from the symmetry of the network with respect to the horizontal and vertical planes indicated by the dotted lines. Fig. 1.10 omits one of the stable states, namely state **0**, which is always stable in a network of McCulloch-Pitts neurons with no thresholds or biases.

Table 1.1 compares the present "WXZ notation" to the "TVI notation" used consistently by Hopfield (1982 and subsequently).

Table 1.1 Terminology and Notation

Terminology	Notation	
	WXZ	TVI
Strength of the connection from the jth unit to the ith, called the weight	Wij	T_{ij}
Activation or output of unit i	X_i	V_i
Externally-applied input to i^{th} unit	Z_i	I_i
Net input to i from all other units	$Y_i = \Sigma_j W_{ij} X_j$	
Internal threshold (bias) of unit i	Θ_i	U_i
Total input to i^{th} unit	$U_i = Y_i + Z_i$	$= u_i$
Activation (firing rate function)	F	f
I/O relation	$X=F(U-\Theta)$	$V=f(u)$
Network state	$\mathbf{X}=(X_1,..X_N)$	$V=(V_1,..V_N)$
Computational energy	$H(\mathbf{X})$	$E(\mathbf{V})$

Hopfield (1982) noted three essential differences between *perceptrons and collective computation networks*:

(1) Perceptrons were modeled chiefly with neural connections in a forward direction. All interesting properties of collective ... computation networks arise from *strong, nonlinear feedback*.

(2) Perceptron studies usually made a random net deal directly with a stimulus world and did not ask the questions essential to finding the more abstract *emergent computational properties*.

(3) Perceptron modeling used synchronous neurons like a conventional digital computer. "Chiefly computational properties which can exist *in spite of asynchrony* have interesting implications in biology."

The first distinction is neatly captured in Lippmann's (1987) diagram of the Hopfield net (fig. 1.11). A binary input **x**(0) is applied at time zero and "the net then iterates until convergence" to a stable state $\mathbf{x}^o$, which satisfies the fixed point equation (6). The next chapter is essentially devoted to Hopfield's third point. We shall see that the emergent computational properties — arising from the strong, nonlinear feedback — exist *because of asynchrony*, the meaning of which can hardly be stated too clearly.

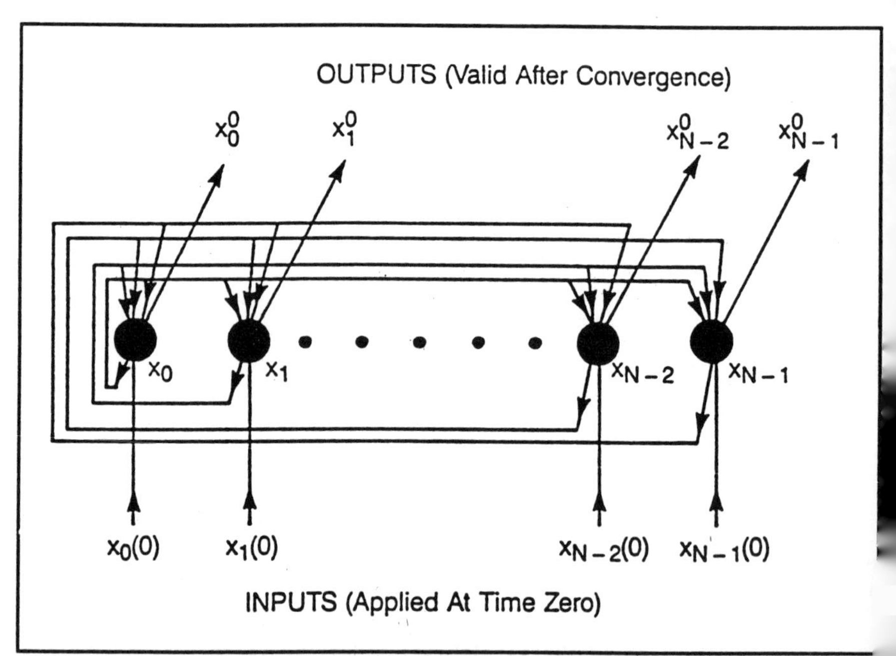

Figure 1.11 An unknown binary input pattern is applied at time zero and the net then iterates until convergence when node outputs remain unchanged. The output is that pattern produced by node outputs after convergence.

Additional Problems

1.6 What elementary logic functions are computed by this network of McCulloch-Pitts neurons? In other words, classify the truth tables D(A,B), E(A,B), and F(A,B) corresponding to the three output units.

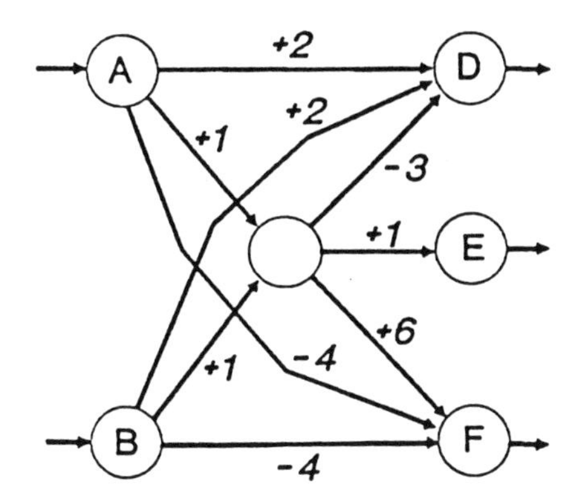

1.7 Exhibit a set of weights that makes this network of four McCulloch-Pitts neurons, with no thresholds or biases, as shown below, solve exclusive-OR problems. The additional requirement is that the interchange of weights W_{32}= a and W_{31}= b does not preserve the logic.

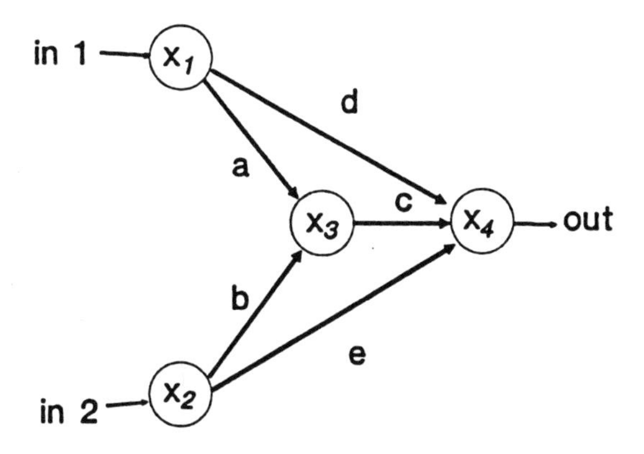

1.8 The outer product rule is used to construct a two-layer associative memory network in which the n-bit output patterns all have Kronecker δ's for their components.

$$b^k = (\delta_{k1},\ldots,\delta_{kn})$$

where k is the index of the corresponding input pattern. Assume McCulloch-Pitts neurons. How does this procedure differ from Rosenblatt's prescription for constructing a "universal perceptron?"

Solutions:

1.6 AND, OR, XOR.

1.7 (a,b,c,d,e) = (1,-2,2,-1,.5)

2. Three Faces Of Neurodynamics

In addition to a function that prescribes the manner in which neurons are to fire, we will need a system that describes the manner in which the state of each neuron is updated. Such a system is referred to as the dynamics of the neural net or simply, the neurodynamics. Immediately we arrive at a division into two systems: a continuous update described by a system of differential equations or a sequential system in which first one neuron is selected for update and then another. In the latter (discrete) case, the method of selection gives rise to two subcases — either the neuron to be updated is chosen in accordance with some pre-described condition or plan, (synchronous) or else it is chosen at random (asynchronous). These are the three faces of neurodynamics: the continuous, the synchronous and the asynchronous (or Glauber) dynamics.

Models of Neural Networks

Hopfield (1984) has modelled the neural network by a collection of RC coupled amplifiers and resistors. The neurons are represented by the amplifiers which, like the old diode tubes of the Geiger counter have only two states - OFF and ON. An amplifier that is producing an output voltage, V, represents a neuron that is firing at a rate, V/V_{max}, relative to its peak rate where V_{max} is the greatest output voltage that the amplifier can produce. The strength of the connection between amplifier i and amplifier j is inversely proportional to the magnitude of the coupling resistance:

$$T_{ij} \propto 1/R_{ij}$$

In order to allow for inhibitory coupling, each amplifier is paired with an invertor so that either the positive or the negative voltage is taken.

The voltages involved are (1) an output voltage from each neuron, V_i, and (2) an input voltage to each neuron, u_i. The two are related by the characteristic operating curve of the amplifier: $V_i = F(u_i)$. The input to amplifier i is a suitable linear combination of the output voltages of all other amplifiers (neurons). As the output voltages vary, so too do the inputs and, following Hopfield, we have a general system:

$$C \frac{du_i}{dt} = \sum_{j=1}^{N} T_{ij} F_j(u_j) - \frac{u_i}{R} + J_i \qquad \textbf{(1)}$$

where u_i is the input voltage of the ith amplifier in the circuit, C is the capacitance, and R the resistance associated with each neuron. The matrix is the conductance (reciprocal of the resistance) associated with the path from neuron j to neuron i and F is the activation function of the jth neuron ($0 \leq F(u_j) \leq 1$). J_i is the result of an externally applied current. This system is a special case of the system of differential equations considered by Cohen and Grossberg (1983).

If there is only one neuron in the circuit, then:

$$C \frac{du_1}{dt} = -\frac{u_1}{R} + J_1 \qquad \textbf{(2)}$$

(Since $T_{11} = 0$). This equation is readily integrated to get:

$$u_1(t)\, e^{\frac{t}{RC}} - u_1(0) = \frac{1}{C} \int_0^t J_1(\tau) e^{\frac{\tau}{RC}}\, d\tau \qquad \textbf{(3)}$$

The correctness of this is easily seen by applying Kirchoff's laws to the circuit of figure 2.1. I_R and I_C must add up to I_{bias} and V_R AND V_C must be equal (to u_1). Since $I_C = C\, du/dt$ and $I_R R = V_R = u$, we immediately have:

$$C\frac{du_1}{dt} = -\frac{u_1}{R} + I_{bias} \qquad \textbf{(4)}$$

With the input to the neuron, $J_1 = I_{bias}$, this is the same as equation (2).

If we start with $u_1 = -A$ and take I_{bias} to be a step function at time t* with height A/R, then, from equation (3):

$$u_1\, e^{\frac{t}{RC}} + A = \frac{1}{C} \int_0^t I_{bias}\, e^{\frac{\tau}{RC}}\, d\tau$$

$$u_1\, e^{\frac{t}{RC}} + A = \frac{1}{C} \int_0^t \frac{A}{R}\, \mathcal{H}(\tau - t^*)\, e^{\frac{\tau}{RC}}\, d\tau \qquad \textbf{(5)}$$

$$u_1 = A\, [1 - e^{-\frac{t-t^*}{RC}}] - A\, e^{-\frac{t}{RC}} \qquad (for\ t > t^*)$$

The state of the neuron is $x_1 = \mathcal{H}(u_1) = \mathcal{H}(t-t')$ where t' is the solution to the equation $u_1 = 0$.

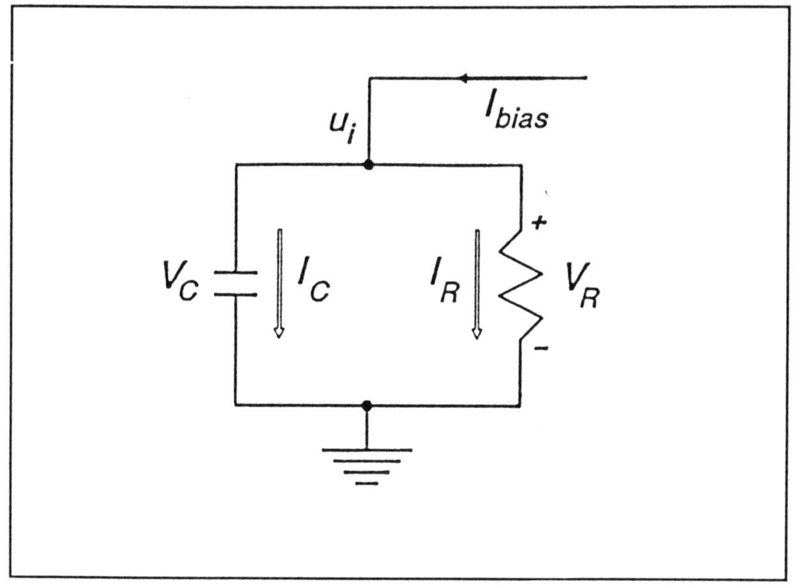

Figure 2.1 One-neuron circuit diagram.

Problem 2.1 : If $t^* = 0$, show that: $t' = RC\ \ln(2)$.
Solution:

$$A[1 - e^{\frac{-t'}{RC}}] - A\ e^{\frac{-t'}{RC}} = 0$$

$$1 - 2e^{\frac{-t'}{RC}} = 0$$

$$\frac{t'}{RC} = \ln 2 \quad .$$

In solving a more complicated problem involving many neurons, we will be immediately confronted not with one, but with an entire system of differential equations.

Continuous Neurodynamics - The First Face

Here we are dealing with a system of nonlinear differential equations and, as everyone who has taken the graduate course in differential equations knows, the odds are heavily against being able to find solutions to these equations. What we seek is not so much a description of the evolution of the system in time as the equilibrium points where the system comes to rest. Suppose we are dealing with a system of N neurons and the firing state of each neuron is described by real numbers $y_i \in [-1,1]$. The collection of all states is summarized by a vector, **y**, and the evolution of the collection of neurons can then be described by:

$$y' = f(y) \tag{6}$$

where **f** is a suitably chosen vector valued function. In technical terms, this is a first order autonomous system of equations. If **f(y)** is not zero, then **y**$'$ is not either, and **y** continues to change. The only place where the system can come to rest is a point in E^n where $\mathbf{f}(\mathbf{y}) = \mathbf{0}$. Such a point is called an equilibrium point; and there are three kinds: stable, unstable and neutral. They can be visualized in terms of a simple cone. A cone balanced on its base is as good an example of a stable equilibrium as we could ask for — at least if we are only concerned with tipping it over. If the cone axis is rotated ever so slightly with respect to the vertical, forces arise to right the cone and render the axis vertical. On the other hand, if the cone is balanced on its vertex, there will be an equilibrium because a cone *can* be balanced in such a position, but the equilibrium is unstable. If the axis of the cone deviates ever so slightly from the vertical, forces arise that tend to push the axis further from the vertical. This is a good example of an unstable equilibrium. Notice, it is not possible to balance the cone with the axis not vertical. These positions are not equilibrium positions.

Returning to the example of the cone balanced on its base, if the cone is not tipped, but displaced, (i.e., moved horizontally on the table) there will be no forces tending to either restore the cone to its original position (not stable) or to move it further away (not unstable). This equilibrium is called neutral.

A neural network has points at which it can be "balanced" — points where the system will remain if it arrives there. These equilibrium points can be classified as stable or unstable depending on whether, when the network finds itself in a "nearby" state, it moves closer to the equilibrium or further from it.

Problem 2.2 : Find the equilibrium points for the following system of equations:

$$x' = x - y^3$$
$$y' = y - x^3.$$

Solution: (0,0), (1,1) and (-1,-1).

Tests for Stability

If we assume that the function **f** of equation (1) has a few derivatives, we can use the vector form of Taylor's theorem to describe the nearby states. Let $\mathbf{y}_0$ be an equilibrium point and let $\mathbf{z} = \mathbf{y} - \mathbf{y}_0$. Then (6) can be rewritten as:

$$z' = f(y_0) + f'(y_0)z + \tfrac{1}{2} z^T f''(y_0)z + \ldots \quad \textbf{(7)}$$

where $\mathbf{f}'(\mathbf{y}_0)$ is the matrix of first partial derivatives of the components of **f** and $\mathbf{f}''(\mathbf{y}_0)$ is a three-dimensional "matrix" of second partial derivatives of the components of **f**. Since $\mathbf{y}_0$ is an equilibrium point, $\mathbf{f}(\mathbf{y}_o) = \mathbf{0}$ and, if we let **A** be the matrix $\mathbf{f}'(\mathbf{y}_0)$, then:

$$z' = Az + R(y_0,z) \quad \textbf{(8)}$$

where **R** represents the sum of the rest of the series. Under reasonably mild restrictions on **f**, as $\|\mathbf{z}\| \to 0$, $\|\mathbf{R}\|/\|\mathbf{z}\| \to 0$.

A great deal of information can be obtained concerning the nature of the equilibrium point $\mathbf{y}_0$ ($\mathbf{z} = 0$) by examining the matrix **A**. If **R** is not big enough to interfere, then **A** will determine the stability of the equilibrium. In particular, if **R** is small and the eigenvalues of **A** have strictly negative real parts, then $\mathbf{z} = 0$ is a stable equilibrium. If one or more eigenvalues of **A** have positive real parts, then the equilibrium tends to be unstable. If **A** has a zero eigenvalue, the nature of the equilibrium is determined by **R**.

The question of how small **R** has to be is dealt with exhaustively in authoritative works such as Bellman (1953), Hirsch and Smale (1974) and Hartman (1964).

Problem 2.3 : The origin is an equilibrium point for the system

$$x' = x - y$$
$$y' = 2x + 3y .$$

Is it stable?

Solution:

$$W = \begin{pmatrix} 1 & -1 \\ 2 & 3 \end{pmatrix}$$

$$\det(W - \lambda I) = \lambda^2 - 4\lambda + 5 = (\lambda - 2)^2 + 1.$$

The eigenvalues are 2 + i and 2 - i. The real part is positive, so the origin is not a stable equilibrium point.

Lyapunov Methods

To prove the stability of a solution to a system of differential equations, Lyapunov introduced the function that bears his name. The Lyapunov function, V for our system of equations has the following properties:

1. It has continuous partial derivatives with respect to each y_i;
2. $V(\mathbf{y}) \geq 0$ when $\|\mathbf{y}\| \geq 0$;
3. $\nabla V(\mathbf{y}) * \mathbf{f}(\mathbf{y}) \leq 0$.

The significance of this can be seen from the solution curve of (1) represented by the dashed line in fig. 2.2. See Hartman (1964).

The level line of the Lyapunov function is at right angles to the gradient which, as the reader undoubtedly knows, points in the direction of most rapid increase of the function. The tangent line to the solution curve of (1) makes an angle θ with the level line and, if that angle lies between 90° and 270°, (and it will if $\cos\theta < 0$) then the curve is always moving in toward the equilibrium point and never out and away from it. Such an equilibrium must be stable. Moreover:

$$\begin{aligned} \frac{dV}{dt} &= \nabla V(y) \cdot y'(t) \\ &= \nabla V \cdot f(y) \\ &= \|\nabla V\| * \|f(y)\| * \cos(\theta) \leq 0 \end{aligned}$$

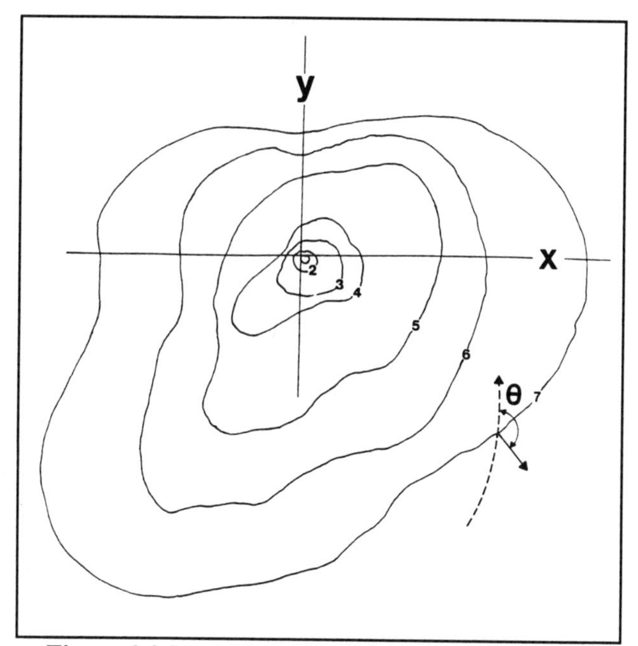

Figure 2.2 Level lines of the Lyapunov function.

ınd, since the last quantity is negative, θ must be in the second (or third) ɪuadrant. Hence if there is an equilibrium point, $\mathbf{y}_0$, where **f** vanishes and if here is a function, **V**, satisfying the three conditions above, then $\mathbf{y}_0$ is a stable equilibrium point.

Global Stability

The reader familiar with differential equations is cautioned against using ıe conventional definition of this term (Hartman 1964, 537). In neural etworks, global stability refers to the network rather than the individual solution. et us observe that the solution of a system of differential equations such as (1) ould, after a long time, get closer and closer to a single point. (Which point ay well depend on where the system started.)

A second possibility is that the solution ultimately gets closer and closer a closed curve in E^n. This is called the orbit of the solution. Still a third ssibility is that the solution jumps back and forth between some collection of ints in a chaotic way. (Chaos is a relatively recent discovery in the field of differential equations, although its roots extend back quite a way. We will give an example of it when we discuss synchronous dynamics.)

These last possibilities we group together under the heading of oscillations and they are undesirable in a neural network. They indicate that the neural network cannot make up its mind. (This description is not mere personification.)

We define a neural network as being globally asymptotically stable if,

a. regardless of the starting point, the system tends to an equilibrium point,
b. that point is isolated from all others, and
c. the point must remain practically the same if the parameters are varied slightly.

See Cohen and Grossberg (1983) and Grossberg (1978). It is known that the now-classical perceptron of Frank Rosenblatt with its feed-forward-only dynamics is stable. But the introduction of feedback is a destabilizing influence. The work of Cohen and Grossberg has shown that: if the connectivity matrix, **W**, is symmetric (as in the Hopfield net to which much of this book is devoted), then the system is stable. The asymmetric case is not quite so tractable, and unstable networks have been found. Indeed chaotic networks seem to exist. Under reasonably mild restrictions, Cohen and Grossberg (1983) show that the usual models for neural networks result in global asymptotic stability. There are a finite set of isolated equilibrium points and these correspond to the answers that we wish to get from the system.

<u>Problem 2.4</u> : Compute the Lyapunov function for the system

$$x' = -x$$
$$y' = -y$$

which has an equilibrium point at the origin. Characterize the equilibrium as stable or unstable.

Solution:

$$V(x,y) = \tfrac{1}{2}x^2 + \tfrac{1}{2}y^2 .$$

The existence of the Lyapunov function implies that the equilibrium is stable.

Problem 2.5 : Find a Lyapunov function for the system:

$$\frac{dx}{dt} = -x - 2y + x^2y^2$$

$$\frac{dy}{dt} = x - ½ y - ½ x^3y$$

Solution: It is often useful to try $V = ax^2 + by^2$.

$$\frac{dV}{dt} = 2ax\left[-x - 2y + x^2y^2\right] + 2by[x - ½ y - ½ x^3y]$$

$$= -2ax^2 - 4axy + 2ax^3y^2 + 2bxy - by^2 - bx^3y^2.$$

et $b = 2a$ and:

$$\frac{dV}{dt} = -2ax^2 - 2ay^2 = -2a(x^2 + y^2) < 0.$$

Synchronous Dynamics - The Second Face

In order to implement a continuous neurodynamics in any practical etwork, the differential equations to which the dynamics gives rise must be ıtegrated numerically. The differential equations are then replaced by ifference equations and (1) becomes:

$$y(n+1) = y(n) + \Delta t \;\; f(y(n))$$

'here each component of the right-hand side is evaluated and then used to pdate **y**. It is but a short step to generalize this to:

$$\mathbf{y}(t+1) = \mathbf{g}(\mathbf{y}(t)) \qquad (9)$$

'ith the understanding that *each* new component of **y** be computed before rogressing to **y**(t+2). We might as well order the components of **y** in the rder in which they will be updated. Hence:

$$y_i(t+1) = g_i(y(t))$$

for i = 1,2,3,...n. To relate all this to biological neurons, we can suppose th a global clock ticks off time at t = 0,1,2, ... The neurons are a synchronized and update their current states at each tick of the clock. The stat are updated serially. The time between updates should not be small compare to the characteristic times of the system, or the synchronous dynamics will lo its separate identity and become a mere approximation to the continuo dynamics.

The stability of this process has been investigated by Marcus ar Westervelt (1990) and they concluded that at least in the case:

$$y^{n+1} = F(W\, y^{n} + z) \qquad \textbf{(10)}$$

where **W** is symmetric and $\beta = \sup|F'(p)|$ satisfies the inequality:

$$\frac{1}{\beta} > -\lambda_{\text{min}} \qquad \textbf{(11)}$$

where λ min is the smallest eigen value of W, that the synchronous dynami is stable.

On the other hand, here, as in the continuous case, asymmetr connectivity matrices (**W**) can give rise to periodic orbits and even to chaos.

So far we have treated the synchronous update model — the second fa of neurodynamics — as an aberration that results from either the biologicall implausible assumption of a global clock or the mathematically indefensible us of an oversized time step in going from the differential equations of the neur circuit to a finite difference formalism. Does this dynamical option have r interesting features at all?

The answer, happily, is that it does, for reasons that Marcus, Waugl and Westervelt (1990) have explained with ample clarity in their discussion o associative memory in an analog iterated map neural network. To full appreciate their work, however, the student might require a prior understandin of the issues covered in the next two chapters in relation to the Hopfield (198 1984) models. These issues include the global stability of neural networks wi symmetric weight matrices (Cohen and Grossberg 1983) and the corresponden between fixed point attractors and the network states of minimum energy. A the risk of introducing chaos into our own exposition of these issues, this sectio will conclude by sketching what the Harvard group discovered about th "iterated map" neural network. The reader who has trouble with this sketch welcome to return to it after completing the remaining text which, except fo several cautions and asides, will focus on the continuous, deterministic dynami of neural circuits and the asynchronous (Glauber) dynamics of Hopfield original stochastic model.

The Iterated Map Neural Network

Consider a map of the form:

$$\boldsymbol{x}(t+1) = \boldsymbol{F}[\boldsymbol{x}(t)], \quad t = 0,1,2,... \tag{12}$$

where $\mathbf{x}(t) \in E^2\ \forall t$. The goal is to describe the attractors of the system. Let the mapping **F** be defined by:

$$\begin{aligned} x(t+1) &= c - \tfrac{1}{2} x^2(t) + y(t) \\ y(t+1) &= \tfrac{1}{2} x(t) \end{aligned} \tag{13}$$

where c is an arbitrary constant. Setting c = 0.99, and starting from just about any initial point $\mathbf{x}(0) = (x(0), y(0))$, the system settles down to a limit cycle in which the sequence $\{\mathbf{x}(t): t = 0,1,2, ... \}$ oscillates between points (1.6, -.3) and (-.6, 0.8), approximately. This limit cycle is the attractor of the system.

On the other hand, if c = 2.133, the behavior of **x**(t), as t goes to infinity, is not periodic but chaotic, meaning (loosely speaking) that the path it follows is superficially unpredictable or random — even though it comes *from a strictly deterministic system.* The following BASIC program plots ten thousand points (line 180) of the chaotic series in that portion of the (x,y)-plane which lies inside the frame drawn by lines 10 through 60. Lines 100 and 110 define **x**(0) and the loop beginning on 130 iterates the mapping. The result is a "strange attractor" — a constellation of points toward which the state vector **x** is inexorably drawn from any starting point on the background of empty space (fig. 2.3).

```
10 Xmax = 4
   Ymax = 2
   SCREEN 9
   WINDOW (-2*Xmax, -1.4*Ymax)-(1.1*Xmax, 1.1*Ymax)
   LINE (-Xmax, -Ymax)-(Xmax, Ymax), 0, BF
60 LINE (-Xmax, -Ymax)-(Xmax, Ymax), , \b
100 x1 = .5
110 y1 = 0
120 c = 2.133
130 FOR n = 1 TO 10000
```

```
140      x = c - .5 * x1 ^ 2 + y1
150      y = .5 * x1
160      x1 = x
170      y1 = y
180      LINE (x1, y1)-(x1 + .0001, y1 + .0001)
190 NEXT n
   END
```

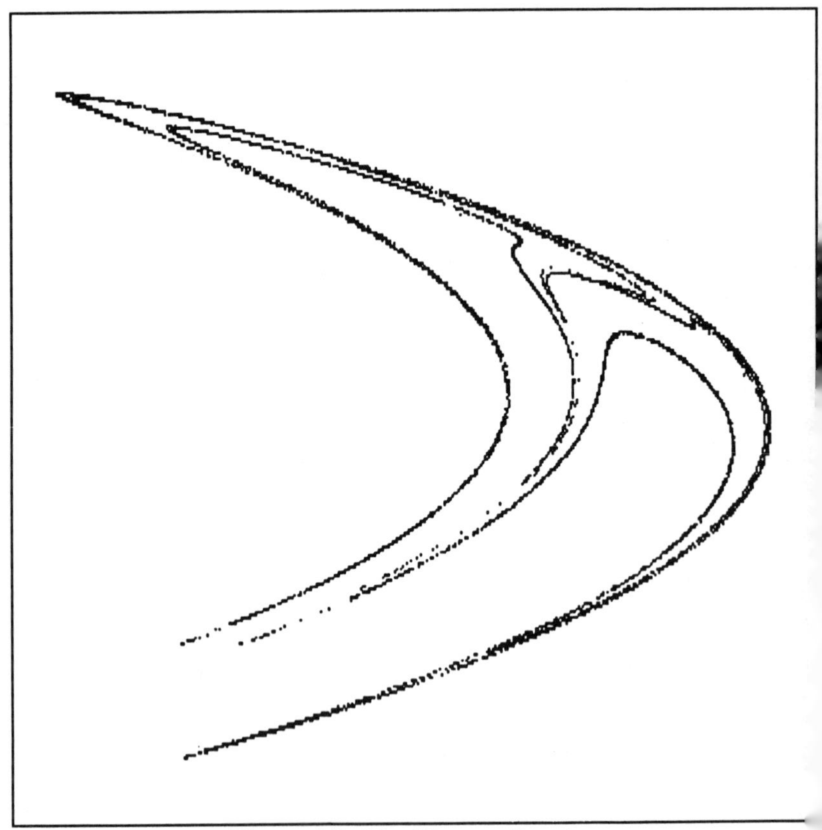

Figure 2.3 The Henon attractor.

Designing the Neural Network

The design of a neural network associative memory is the inverse of this problem type: The starting point is the set of desired attractors and the problem is to find a dynamical system that possesses these attractors and, to the extent possible, no other "spurious" attractors. (Recall Hopfield's explanation, which was quoted in chapter 1.) Moreover, the attractors of the neural network should be fixed points instead of limit cycles or the strange attractors found in chaotic systems. The fixed points correspond to states to which the network must converge if it is to be useful. The iterated map neural network follows equation 10, which is repeated here:

$$y^{n+1} = F(W\, y^{n} + z)$$

In terms of the components, this can be written:

$$x_i(t+1) = F_i\left\{\sum_{j=1}^{N} W_{ij}\, x_j(t) + z_i\right\}, \qquad i = 1,2,\ ...\ ,N \quad . \qquad \textbf{(14)}$$

The fixed points of the system so described, which are indicated by a superscript, 0, all have the property that:

$$x^0 = F[Wx^0 + z] \quad , \qquad \textbf{(15)}$$

i.e.:

$$x_i^0 = F_i\left\{\sum_{j=1}^{N} W_{ij}\, x_j^0 + z_i\right\}, \qquad i = 1,2,\ ...\ ,N$$

so that, once the sequence $\mathbf{x}(t)$ hits an $\mathbf{x}^0$, it will remain fixed at those coordinates. The network (or system) is globally stable if, starting from an arbitrary $\mathbf{x}(0)$, it eventually ends up at a fixed point.

Must the iterated map neural network converge to fixed points? The answer depends on the weight matrix **W** and on the nonlinear activation functions $F_i(.)$ of the N neurons. Using the McCulloch-Pitts neurons, so that the $F_i(u) = \mathcal{H}(u)$ for every i, the fixed points correspond to stable states defined by vectors of N binary-valued components. In this case the attractors may be either fixed points or period-two limit cycles (as in the example system above with c = .99). The same applies to systems of reciprocally interacting Ising spins in which $F_i(u) = \text{sgn}(u)$. Marcus and Westervelt (1989) have shown that the limit cycles can be eliminated by smoothing out the sharp transitions of the McCulloch-Pitts neurons and Ising spins, i.e., by going from two-state neurons to graded ("analog") neurons. While this is easily accomplished with functional forms like $1/(1+e^{-bu})$ and tanh(bu), there is, in addition to symmetry, a stricter stability criterion that these investigators have derived. In order to converge exclusively to fixed points, the eigenvalues of **W** in the iterated function system (10) must satisfy:

$$\frac{1}{\beta_i} > -\lambda_{min} \quad \textit{for all } i$$

where β_i (> 0) is the maximum slope of $F_i(u)$ and λ_{min} is the minimum eigenvalue of the weight matrix **W**. When F(u) = tanh(bu), as in fig. 2.4a, the maximum slope $\beta = b/2$ is at the origin. The result applies, however, to any monotone increasing activation function as suggested by fig. 2.4b.

Problem 2.6 : Let the three units of an iterated map neural network with:

$$\mathbf{W} = \begin{bmatrix} 0 & +1 & -1 \\ +1 & 0 & +1 \\ -1 & +1 & 0 \end{bmatrix}$$

have the activation function tanh(bu) in common. What upper bound is imposed on b by the requirement of global stability?

Solution: We form **W** - λ**I** and take the determinant:

$$det \begin{bmatrix} -\lambda & 1 & -1 \\ 1 & -\lambda & +1 \\ -1 & 1 & -\lambda \end{bmatrix} = -\lambda \begin{bmatrix} -\lambda & +1 \\ 1 & -\lambda \end{bmatrix} - 1 \begin{bmatrix} 1 & -1 \\ 1 & -\lambda \end{bmatrix} - 1 \begin{bmatrix} 1 & -1 \\ -\lambda & +1 \end{bmatrix}$$

$$= -\lambda(\lambda^2 - 1) - (-\lambda + 1) - (+1 - \lambda)$$

$$= -(\lambda - 1)^2 (\lambda + 2)$$

and the eigenvalues are 1 and -2. For stability,

$$1/\beta > -(-2) = 2 \text{ or } \beta < 1/2.$$

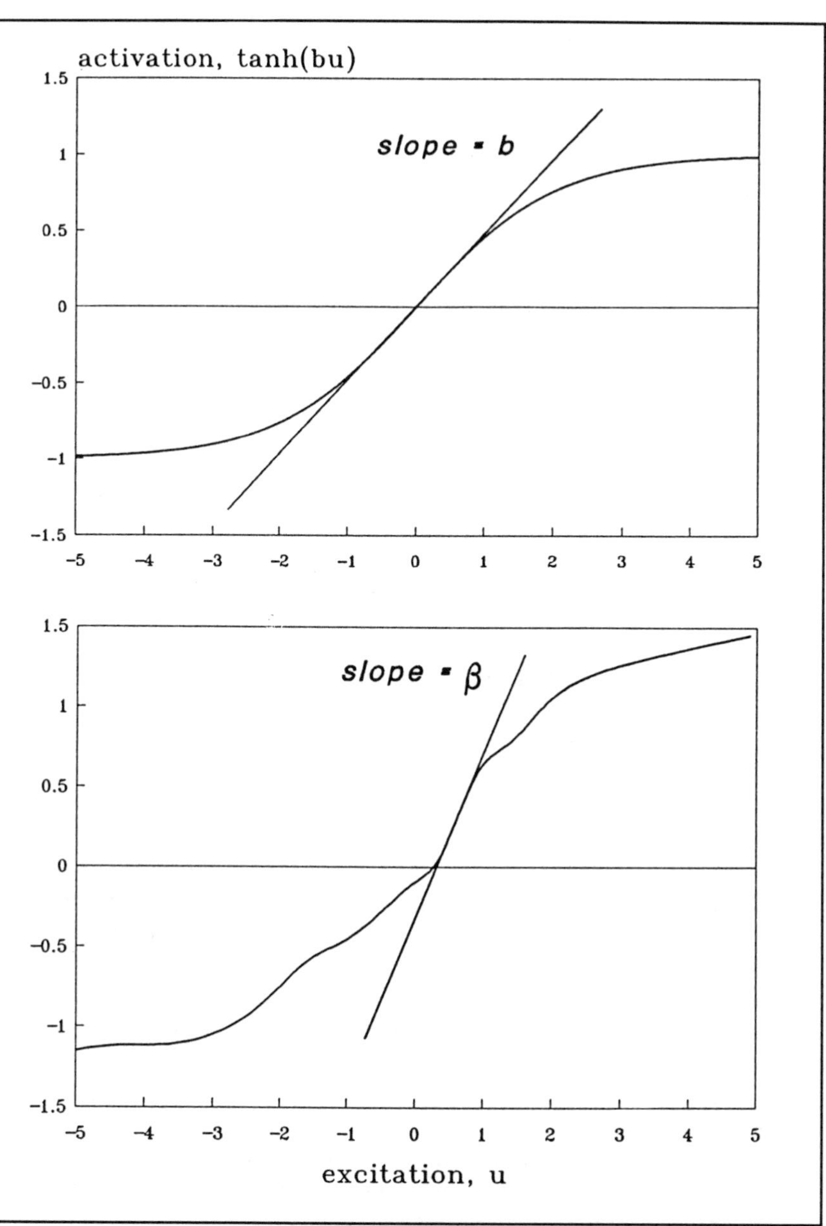

Figure 2.4 The maximum slope of the activation function is β. When $F(u) = \tanh(bu)$, the maximum slope b occurs at $u = 0$. (top). $F(u)$ can be an arbitrary increasing function (bottom).

If F(u) = tanh(bu), then $F'(u) = \text{sech}^2(bu)*b \leq b$. So $\sup|F'(u)| = b$ and b must be less than 1/2 to insure stability.

When, in addition to the symmetry of **W**, the Marcus-Westervelt stability criterion equation (11) is satisfied, the iterated map neural network has the Lyapunov function:

$$E(x) = -\tfrac{1}{2}\sum_{i,j} W_{ij}\, x_i\, x_j - \sum_i z_i\, x_i + \sum_i G_i(x_i) \quad ,$$

where:

$$G_i(x_i) = \int_0^{x_i} F_i^{-1}(x)\, dx \quad ,$$

essentially similar to the Cohen-Grossberg system. Then E(t) will decrease at each discrete time step and, since it is bounded below, it ultimately must come to rest at a (local) minimum value. Moreover, these minima of E occur at points $\mathbf{x}^0$ which are solutions of the fixed point equations (15).

Chaos in Synchronous Dynamics

On the other hand if the symmetry is violated, it is possible to develop oscillations and even chaos. The discovery of chaos in asymmetric neural networks (Sompolinsky et al. 1988) should not be too surprising. Its appearance in continuous systems has been anticipated, based on simulations of discrete neurodynamics. Hopfield, in probing a specimen net of thirty McCulloch-Pitts neurons with random, asymmetric connectivity, observed "chaotic wandering in a small region of state space" (Hopfield 1982). The dynamical assumptions of Hopfield's stochastic model were the same ones introduced by Glauber (1963) to study the Ising model. (See below: The Third Face of Neurodynamics.) Irrespective of connectivity, the Glauber dynamics is dissipative, giving rise to a Markov chain that is ultimately absorbed into irreducible sets of persistent states. In a more systematic Monte Carlo study, using the spreading activation dynamics in which neurons are updated synchronously at regular intervals, Kürten observed limit cycles with mean periods growing exponentially in network size, and speculated that this might presage chaos in continuous systems of similar connectivity.

At least as important as the discovery of chaos is an understanding of its functional roles in biophysics and neuromorphic cybernetics. What makes the recent disclosure of Sompolinsky, Crisanti, and Sommers (1988) particularly interesting is that the phase is controlled by the gain parameter (β) of the individual neuron firing rate function. In the frozen phase, where the rate may be quasilinear over the principal dynamic range, the system evolves to the origin. As the gain increases, nonzero stable states appear. If the storage ratio is too large, there will be a mixture of the stored states and many stable states that are spurious (the spin glass region). For moderate values of the storage ratio, the stored states will form nicely separated minima (the recall region). Further increases in the gain cause the stable minima to disappear and the network oscillates — sometimes even chaotically. In the wake of Hebb's (1949) enduring contribution, it is customary to regard synaptic organization as the sole source of plasticity in neural networks. Thus the phase in Kürten's (1988) networks is determined by connectivity parameters (which appear in the distribution function that generates the synaptic weights). That the gain parameter might be a control variable opens a new perspective on this now classical doctrine.

Regarding artificial neural systems, the *typicality* of chaos needs clarification. Pineda (1987) prefaced a learning algorithm with the caveat that *most* asymmetric networks converge to nonzero stable states from *most* initial states. The dynamical mean field theory of Sompolinsky, Crisanti, and Sommers (1988) solves an ordinary differential equation for local field autocorrelations between two points along the flow. Their results imply that the flow is chaotic; but what does this flow describe? Crisanti and Sompolinsky (1988), using the Glauber dynamics, found (for $N > 100$) that networks with moderate indices of asymmetry converged in more than 90% of all trials. An ensemble of such networks, started at the same time, would collectively exhibit a transient response (as most converged) followed by a steady state in which a relatively few "survivors" cycle among persistent states in the manner of a homogeneous, irreducible, aperiodic Markov chain. Following Hopfield, this random behavior would seem the discrete counterpart of chaos. Does chaos likewise occur in a clear minority of continuous systems?

Storage in Synchronous Dynamics

How many fixed points does the network have? Can they be made to correspond to certain "patterns" that we want to store and recall? Again the answers depend on the activation functions. The overall picture that emerges from the work of Waugh, Marcus, and Westervelt (1990) shows an N-dimensional "landscape" in which the "elevation" at point $\mathbf{x}$ is defined as the value $E(\mathbf{x})$ of the Lyapunov function. The landscape features some finite number of broader, deeper valleys, the "basins of attraction" that (ideally) surround the desired patterns all of which will coincide with some $\mathbf{x}^0$. The activation vector $\mathbf{x}(t)$ is drawn toward one of these objectives just as, in the space of our experience, a ball rolling under the influence of gravity upon a complicated surface will be drawn to the lowermost points. But the ball is also likely to get stuck as soon as it encounters a crevice or pit and never reach those lowermost points. Similarly, the local minima of E(**x**) may be numerous and shallow and the activation vector may well be trapped in one of these shallow minima. Decreasing the steepness of the neuron activation function reduces the number of these spurious fixed points thus making it easier for **x**(t) to reach deeper minima of the Lyapunov function as suggested by fig. 2.5.

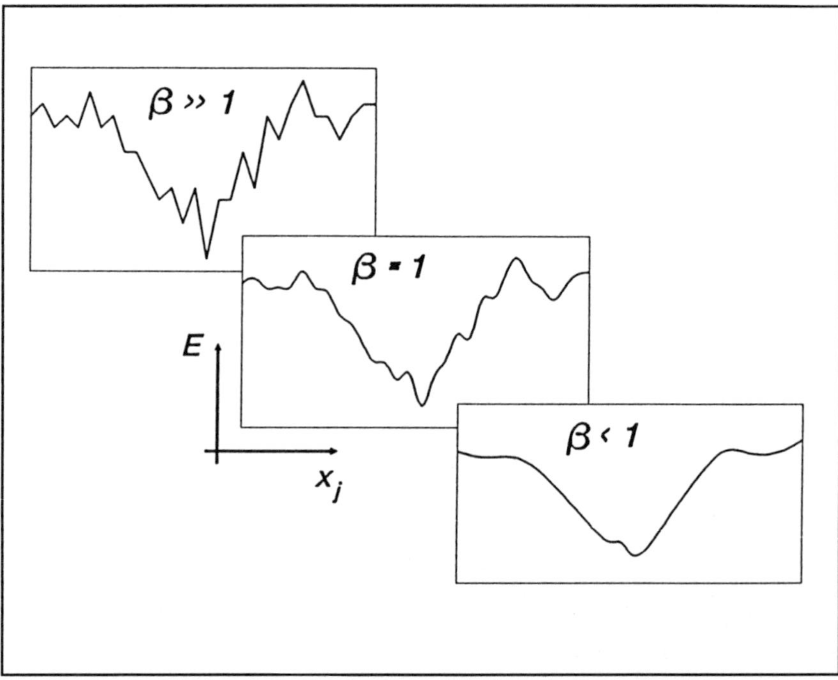

Figure 2.5 Schematic landscapes showing the Lyapunov (energy) function of a state space coordinate. Reducing the steepness (β) of the activation function has a smoothing effect.

This does not mean that perfect recall can be achieved by the iterated map neural network associative memory just by reducing the steepness of the activation functions. Indeed, if the "analog gain" parameter b in the function tanh(bu) is made too small, the network becomes incapable of exhibiting the strongly nonlinear behavior that Hopfield (1982) singles out as the key to making hard decisions. Fig. 2.6 is a "phase diagram" that shows how the behavior of the network varies with position in a plane whose coordinates are the analog gain and the quotient M/N, where M is the number of patterns stored by use of Hopfield's outer product rule. (This rule is adapted from the Hebbian notion of synaptic reinforcement as noted in chapter 1.) If the gain is much below unity, the landscape E(**x**) is dominated by a single, broad basin of attraction about **x** = **0**. When the gain is increased, the behavior of the network depends on the storage ratio: Too many patterns (in the "glassy" region) and the network tends to find a local minimum that is spurious. Settling for a smaller storage ratio can place the operating point inside the "recall" region where stored patterns are recalled reliably. Further increasing the gain moves the operating point into the "oscillation" region where the period-two limit cycles predominate (unless the storage is very small).

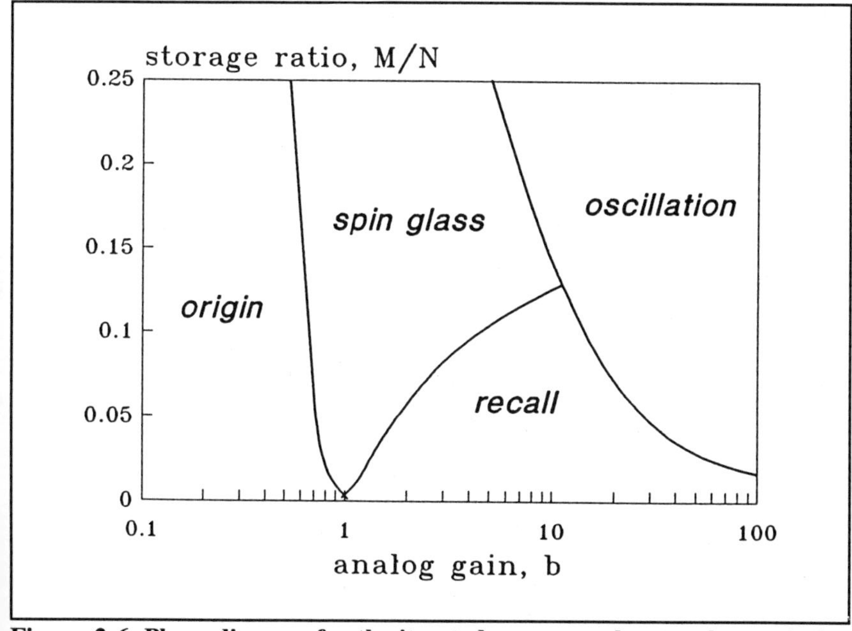

Figure 2.6 Phase diagram for the iterated map neural network associative memory using the (Hebbian) outer product storage prescription and assuming F(z)=tanh(bz). These results apply in the limit of large N.

Phase diagrams are ubiquitous in thermodynamics where the horizontal axis will typically be labeled "temperature." For instance, the three phases of the substance H_2O — water, ice, and steam — will correspond to separate regions in the temperature-pressure plane of fig. 2.7. For the particular choice of nonlinear transfer function F(u) = tanh(bu), the fixed points of equation (10) correspond to the solutions of a mean field theory for the equilibrium magnetization of an Ising spin system with interaction matrix, **W**, at a temperature 1/b. This too can be displayed as a phase diagram. (We intend to discuss the Boltzmann Machine and simulated annealing in later chapters.) But, here, we make some simple observations. The reader who is confused by this is advised to skip the next paragraph and return to it later.

The analog gain parameter thus plays the role of inverse temperature and the smoothing effect of reducing b is analogous to the effect of heating up the system. In particular, the state space coordinate of a "warmer" neural network will be less likely to get stuck in a spurious minimum of the Lyapunov function. (This insight motivates such innovations as "mean field simulated annealing.") It may be helpful to the reader to think of temperature as a shaking of the system. Following Waugh, Marcus, and Westervelt (1990), we emphasize that the synchronous dynamics of the iterated map neural network is not the same as the asynchronous (Glauber 1963) dynamics of spin systems and Hopfield networks at finite temperature. Chapters 6 and 7 are devoted to the latter.

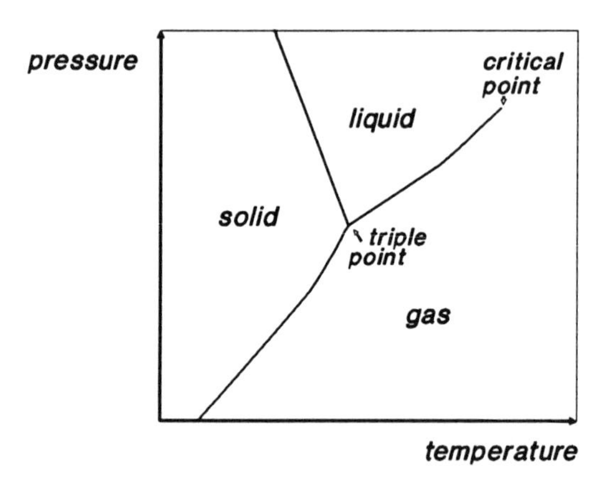

Figure 2.7 Phase diagram for a simple substance.

Glauber Dynamics — The Third Face

The third major system of neurodynamics was introduced by Glauber in order to study phase transitions in the Ising model of ferromagnetism. Briefly put, a neuron is selected at random and the equation:

$$y_i(t+1) = f_i(\mathbf{y}(t))$$

is used to update the state of the neuron selected. Then a (not necessarily[1]) different neuron is selected and the process is repeated. The heart of the Glauber dynamics is its stochastic nature.

Perhaps because the global clock, referred to earlier when discussing synchronous dynamics, lacks a physiological basis or perhaps just to highlight the differences between real (biological) collective computation networks and the artificial ones produced on digital computers, suppose each neuron has its own internal clock which ticks at irregular intervals marking time at $t = 0,1,2,3...$ The only reliable aspect of the process is the probability of the time differences: $T_{k+1} - T_k = H_k$:

$$Pr(H_k < h) = 1 - e^{-h/\Lambda} \qquad \textbf{(18)}$$

where h is positive.

In probability theory, it is proven that the exponential distribution and only the exponential distribution has the property that the time to the next tick is independent of the time since the last tick. Thus the individual neuron clocks are memoryless.

The expected value of $x_i(t)$ is easily computed since $x_i(t) \in \{0,1\}$. Multiply each value of $x_i(t)$ by the probability of getting that value and:

$$Ex_i(t) = 1 * Prob\{x_i(t) = 1\} + 0 * Prob\{x_i(t) = 0\}$$

$$= Prob\{x_i(t) = 1\}$$

$$Ex_i(t) = Prob\{t > t^*\}$$

$$= 1 - e^{-\frac{t}{\Lambda}}$$

by using (18).

In fig. 2.8 we contrast a single-neuron network when confronted by a step excitation using all three different systems of dynamics. For the continuous system, the reader is referred to equations 2.2 - 2.5 and problem 2.1. The input is a step function which begins at time 0 when I_{bias} increases discontinuously

[1]Indeed, it is possible, albeit unlikely, for one neuron to be updated many times before any other neuron is so updated.

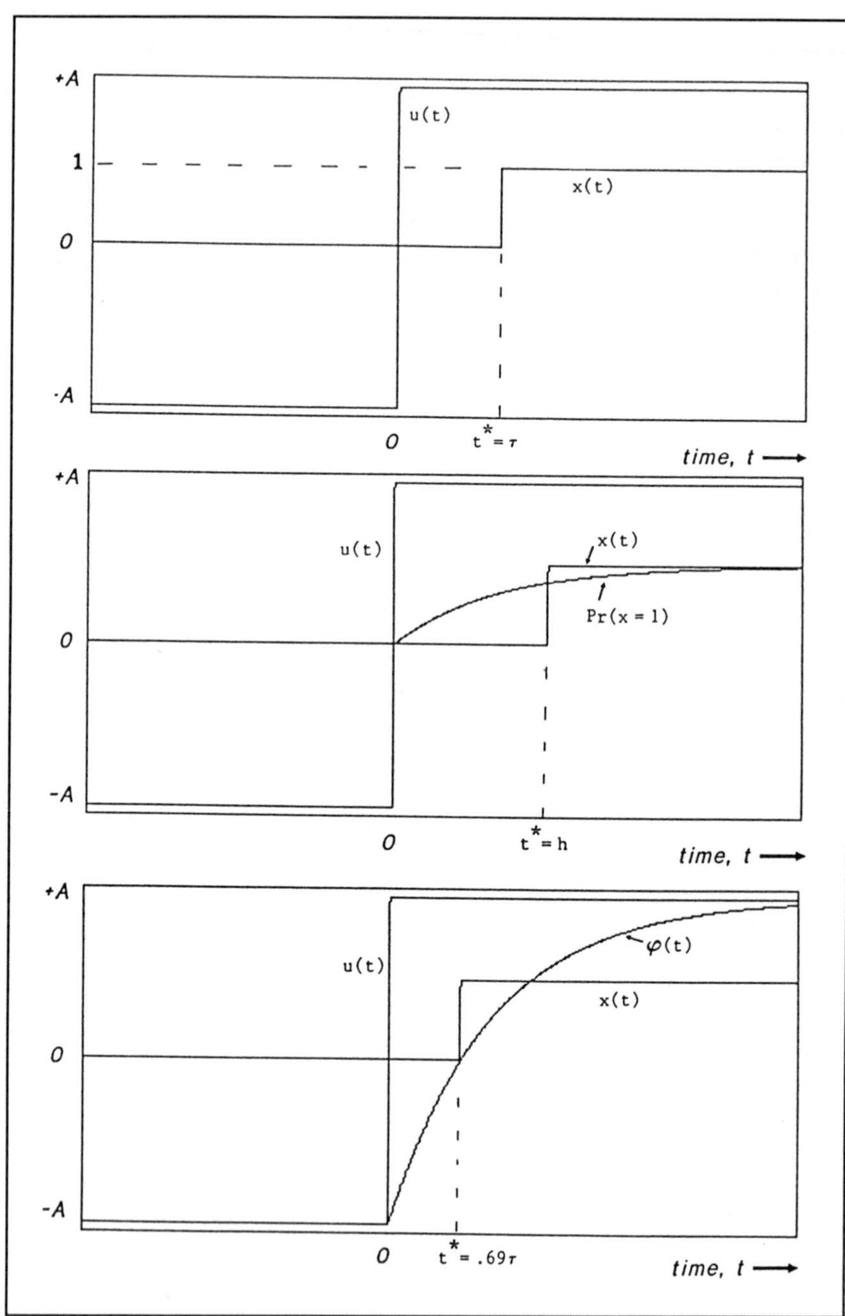

Figure 2.8 Individual neuron responses to step excitation.

rom 0 to A/R: $I_{bias} = \frac{A}{R}\mathcal{H}(t)$. The input voltage is given by equation 2.5 nd $x_1(t)$ = output voltage = $\mathcal{H}(u_1)$. All three are plotted in fig. 2.8c. (See roblem 2.1.)

For the synchronous update or iterated map dynamics, the result is nuch simpler, the bias current turns ON, and the neuron responds (if the current s sufficiently strong). This is shown is fig. 2.8a.

Multineuron Games

A game is characterized by a set of rules governing the behavior of the layers. The rules provide that the game shall consist of a sequence of moves n a particular order; and they prescribe the nature of each move. Following Blackwell and Girshick (1954), the moves are either *personal* or *chance* moves. Each move in chess, for instance, is a personal move. A chance move likewise nvolves a choice among alternatives; but the choice is decided by a random umber generator, e.g., a pair of dice. (Of course, a move may have both lements.)

The following concerns a class of games that evolve in a sequence of istinct *steps* Each step consists of a chance move — the random selection of ne of the players — followed by a personal move, in which the selected player ither (0) quits [drops out] or (1) remains in the game. There are N players, neir interactions specified by an NxN weight matrix **W**. If $i \in \{1,\ldots,N\}$ is the elected player, then i can either quit the game or conduct transactions with the ther active players. The transaction with player j, if $x_j=1$, consists in i anding over to j an amount $|W_{ij}|$, if $W_{ij}<0$; but if $W_{ij}>0$, i receives amount W_{ij} from player j. If i elects to remain in the game ($x_i=1$), the net gain or loss o i will be:

$$y_i = \sum_{j=1}^{N} W_{ij}\, x_j \quad , \tag{20}$$

ith the understanding $W_{ii} = 0$; but if i quits, there are no transactions. Thus x_i is the personal move of the ith player upon selection, the net gain or loss, any, is simply $x_i\, y_i$ at this particular step. It is important to note that the ecision to quit is reversible, since i can reenter the game at the time of next election.

The question of strategy arises when we consider the vector **x** of N inary 0-1 components, which specifies the *state* $(x_i, \ldots, x_N)$ of the game, and sk what the ith player should do if selected at the next step. Evidently there are

2^{N-1} situations that the ith player can face. Strictly speaking, a strategy would define the personal move to be taken by i in response to every one of the contingencies. In order to be practical, the strategies of the N players will need to be expressed in a form more compact than a table of 2^N bits. Consider the strategy:

$$x_i = \begin{cases} 1 \ (''play'') & if \ \ y_i(\mathbf{x}) > 0 \\ 0 \ (''quit'') & otherwise, \end{cases} \qquad \mathbf{(21)}$$

which is $x_i = \mathcal{H}[Y_i]$ in terms of the unit step function $\mathcal{H}[.]$. Note $x_i y_i = x_i \mathcal{H}[x_i] \geq 0$. Thus the effect of strategy (21) is to avert immediate loss. Of course, the decision to remain in the game may lead to net loss up to the time of a player's next selection.

Letting t=0,1,2,... be the number of steps taken from an initial state $\mathbf{x}(0)$ to the present state $\mathbf{x}(t)$, we see that the progress of the game is described by the dynamical equation

$$x_i(t+1) = \begin{cases} \mathcal{H} \left[\sum_{j=1}^{N} W_{ij} \ x_j \ (t) \right] & if \ i=n(t) \\ x_i(t)) & otherwise \end{cases} , \qquad \mathbf{(22)}$$

where $\{n(t), \ t = 0,1,2,\ldots,NT_{max}\}$ is a sequence of independent, random integers, uniformly distributed among the first N. One can imagine spinning a wheel, as in roulette, to pick an n for every step t. This assumes strategy (21) is adhered to by every one of the players.

The sequence ($\mathbf{x}(t)$, t = 0,1,2,...), constitutes a Markov chain. (The probability of reaching a target state from given $\mathbf{x}(t)$, in one step, is affected neither by t nor by the history of game up to the present point.) Given $\mathbf{x}(t) = \mathbf{x}$, there are N+1 alternatives for $\mathbf{x}(t+1)$, since only one player can move at each step. If $\mathbf{x}'$ is one of these alternatives, then the Hamming distance[3] between $\mathbf{x}'$ and $\mathbf{x}$ is no more than one. Let $I(\mathbf{x}' | \mathbf{x}) \ \Sigma \ \{0,1\}$ indicate whether the transition is allowed by strategy (21). Then:

---

[3] $D(\mathbf{x},\mathbf{x}') = \Sigma_i \ |x_i - x_i'|$

$$Pr\ \{\ X(t+1)\ =\ x'\ |\ X(t)\ =\ x\ \}\ =\ (1/N) I(x'|x)\ \textit{if}\ D(x',x)\ \leq\ 1\ , \tag{23}$$

where D(.,.) is the Hamming distance, D(x,y), the factor of 1/N being due to the selection mechanism. The state remains unchanged with probability:

$$Pr\ \{\boldsymbol{X}(t+1)\ =\ \boldsymbol{X}(t)\ \mid\ \boldsymbol{X}(t)=\boldsymbol{x}\ \}\ =\ (1/N) \sum_{x':D=1} I(x|x)\ . \tag{24}$$

Equations (23) and (24) specify all nonzero components of a $2^N x 2^N$ one-step transition matrix that fully characterizes the Markov chain. Moreover, the indicator of accessibility can be written in terms of the states (**x** and **x'**) and the weights as:

$$I(\mathbf{x}'|\mathbf{x})\ =\ \mathcal{H}[\mathbf{d}(\mathbf{x}',\mathbf{x})\mathbf{W}\mathbf{x}^T],$$

where **d**(.,.) has N binary components the ith of which is $|x_i' - x_i|$, **x** is understood as a row vector, and the ^{T} denotes transposition.

Recall (Feller 1957) that each state of a Markov chain is classified as *transient* or *persistent* (recurrent). The closure of a persistent state is an *irreducible* set and its submatrix defines a Markov chain on it that can be treated independently of the rest. Starting from **x**(0), the game passes through a sequence of states thereby taking a random walk on the vertices of a unit hypercube in N dimensions. As $T_{max} \to \infty$, we can inquire into the limiting probabilities of the states, defined as:

$$\pi(x)\ =\ \lim_{t\to\infty}\ Pr\ \{\ X(t)\ =\ x\ \mid\ X(0)\ =\ (1,1,...,1)\ \}\ , \tag{25}$$

where the information given on the right side assumes that all players are in at t=0. This π(**x**) will vanish for all transient states.

If the cardinality of an irreducible, persistent state is unity, the state in question is called *absorbing*. If $\mathbf{x}^0$ is an absorbing state, then:

$$x_i^0\ =\ \mathcal{H}\left[\sum_{j=1}^{N} W_{ij}\ x_j^0\right] \tag{26}$$

for every i. (We set the bias **z** = 0.)

Suppose that **W** is symmetric; $W_{ij} = W_{ji}$ for every i and j. This assumption defines "fair" games in which the interactions are reciprocal; and i gives to (takes from) j exactly the same amount as j gives to (takes from) i. We shall now show that the persistent states of a fair game are all absorbing states. Moreover, these states are one-to-one with the local maxima of a global payoff function:

$$R(\mathbf{x}) = \sum_{i<j} W_{ij}\, x_i\, x_j \quad , \tag{27}$$

in which the notation i<j refers to all (unordered) pairs of players. If player i is selected, the net incentive is given by equation (20) and the decision by equation (21). The payoff to i is:

$$r_i(x\ ;\ x_i) = x_i \sum_{j=1}^{N} W_{ij}\, x_j \quad . \tag{28}$$

The expected value of the payoff, when i is equiprobably any one of the players, is:

$$\begin{aligned} r(\mathbf{x}) &= (1/N) \sum_{i=1}^{N} r_i\ (\mathbf{x}\ ;\ x_i) \\ &= (1/N) \sum_{i=1}^{N} \sum_{j=1}^{N} W_{ij}\, x_i\, x_j \\ &= 2R(\mathbf{x})/N \end{aligned}$$

by the symmetry of **W**.

Consider the change in R due to the move of a player. Let $\Delta x_i = x_i(t\text{-}1)\text{-}x_i(t) \in \{-1, 0, +1\}$. The corresponding change in R is:

$$\begin{aligned} \Delta R(x) &= R(x_1, \ldots, x_i(t+1), \ldots, x_N) \\ &\qquad - R(x_1, \ldots, x_i(t), \ldots, x_N) \\ &= [\ \sum_{j=1}^{N} W_{ij}\, x_j\ ]\ \Delta x_i \quad . \end{aligned}$$

But strategy (21) dictates that the sum in brackets is positive when $\Delta x_i=1$, and negative (or zero) when $\Delta x_i=-1$. Thus $\Delta R \geq 0$; and the sequence ($\mathbf{x}(t)$, t = 0,1,2,...) leads to local maxima of the global payoff function. These local maxima coincide with the coordinates of absorbing states, which satisfy (26), since any move that would depart from such an $\mathbf{x}^0$ would reduce R and be prohibited by the strategy.

This game is formally equivalent to Hopfield's stochastic model, with equations (20) and (21) specifying that the players behave in the same manner as McCulloch-Pitts neurons, and equation (22) stipulating the Glauber dynamics. Hopfield's computational energy (as we shall see subsequently), defined on the domain of (network) state vectors, is just the negative of the global payoff, $R(\mathbf{x})$, of equation (27).

The strategy stated in equations (20) and (21) was a strategy in the strict sense, since it prescribed a move for the selected player given any possible state of the game. Its adoption made the players formally equivalent to McCulloch-Pitts neurons.

A great deal of game theory deals with equilibria, though in most cases the equilibrium strategy is too complex to be employed by real players (Rice 1979). The concept of equilibrium defined by Nash (1951) is one of the central ideas of game theory. Myerson (1978) recalls that:

$$G = \{S_1,\ldots,S_N;r_1,\ldots,r_N\}$$

is a simple N-player game in normal form if each S_i is a nonempty finite set and each r_i is a real-valued function (the reward) whose domain is the Cartesian product of the strategy sets, $S_1xS_2x. . .xS_N = \mathbf{S}$. Interpret {1, . . . N} as the set of players (as before). For each player i, S_i is the set of pure strategies available. Each r_i is the utility function for player i so that $r_i(s_1,. . .,s_N) = r_i(\mathbf{s})$ would be the payoff to i if each player adopts the strategy indicated by equation (21). The Nash equilibrium is reached when no player can unilaterally increase his payoff by switching to a different strategy.

Let $S_i \in \{0,1\}$ for each i so that the pure "strategies" are (0) to quit and (1) to keep playing. The payoff is:

$$r_i\,(\,s\,;\,s_i\,) = s_i \sum_{j=1}^{N} W_{ij}\, s_j$$

in terms of an NxN matrix $\mathbf{W}$, using the "substitution notation" (Nash) on the left hand side. Except for the use of the symbol $\mathbf{s}$ instead of $\mathbf{x}$, this is the same as equation (27). The semantic substitution of "strategy" for "move" leads to the conclusion that the Nash equilibria of G (with symmetric $\mathbf{W}$) are one-to-one with the stable states of the corresponding Hopfield network. These equilibrium points are always reached by the collective activity of players who quit to avoid immediate loss, and keep playing to realize immediate gain.

In the strict sense, the strategic option presents itself to all players simultaneously, at the start of the game, and may not be reversed thereafter. In this sense, the behavior of the McCulloch-Pitts neuron embodies a particular strategy that is time-invariant; and the above results pertain to games and networks in which all players adopt this same strategy. The strategies available to the individual player correspond to the kinds of neurons from which an equivalent network could be constructed. To preserve the notion of a multineuron game, these strategies will have to use only local information: The strategy of player i will use only the information contained in the state vector (**x**) and the N-1 weights on the ith row of **W**.

The objective of designing a multineuron game — to reach suprema of global payoff function (or negative energy) — could lead one to seek *equilibrium points* as N-tuples of strategies (corresponding to neuron types) "such that each player's strategy maximizes his payoff if the strategies of the others are held fixed" (Nash 1951). Then each player's strategy will be optimal against those of the others. Such equilibrium points might correspond to optimal designs of parallel distributed processing machines representing some refinements of the original Hopfield model.

Additional Problems

2.7 Consider this network of four units. Let the units be McCulloch-Pitts neurons so that $x_i \in \{0,1\}$, $i=1,2,3,4$.

(a) What are the stable states of the network? Compare their basins of attraction assuming the synchronous dynamics.

(b) Compare their basins of attraction using the stochastic dynamics and compute the probability of equation (25).

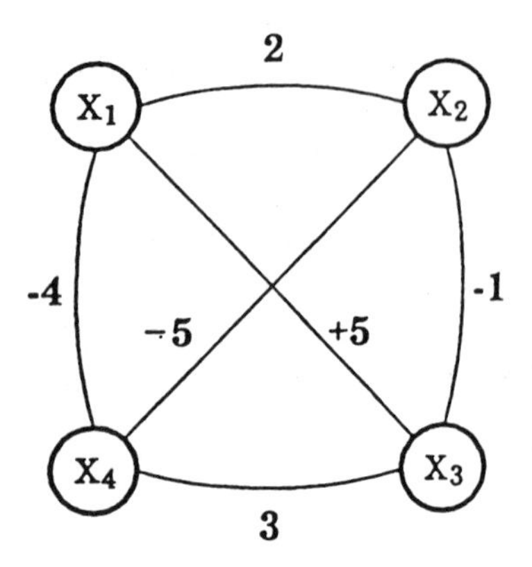

(c) Now let the units be sign neurons (Ising spins), $x_i \in \{-1,+1\}$. Which states are stable? Starting from $(x_1x_2x_3x_4) = (++++)$, does the synchronous dynamics produce convergence?

2.8 Consider this network of three McCulloch-Pitts neurons.

(a) Which states are stable?

(b) Calculate the size of each basin of attraction, defined as the probability of eventual absorption into the stable state given that the initial state is equiprobably any one of the eight possible states, using the stochastic dynamics.

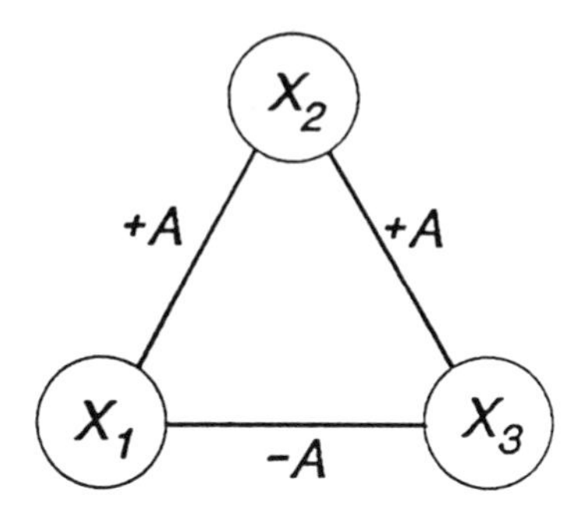

(c) Repeat (a) and (b) assuming sign neurons.

3. Global Pattern Formation and Energy Minimization in Symmetrically Interconnected Networks

While the diversity of neurodynamic assumptions was highlighted in the preceding chapter, their essential unity is expressed by the *fixed point equations*:

$$x_i = F\left(\sum_{j=1}^{N} W_{ij}\, x_j + z_i\right), \quad i=1,\ldots,N \ , \tag{1}$$

whose solutions, written generically as $\mathbf{x}^o = (x_1^o,\ldots,x_N^o)$, are the *stable states* of a network of N neurons with common F. The fixed points are insensitive to the dynamics. Therefore it does not matter how $\mathbf{x}^o$ is arrived at: The state is stable irrespective of the path leading to it.

The synchronous update model (Chapter 2, equation (10)) amounts to iteration of the vector equation $\mathbf{x} = F(\mathbf{Wx} + \mathbf{z})$ until a fixed point is found. In networks of binary units, where F is the step or the sign function, there are 2^N possible states that might be tested for stability. Even for moderate N, this is too much computation. The synchronous update model will often give convergence to stable states in one or two passes. On the other hand, the synchronous dynamics can also fail to converge, becoming trapped into persistent oscillations or limit cycles, despite the existence of stable states guaranteed by the symmetry of the weight matrix.

Therefore we let the neurons "collectively compute" the stable state, given an initial state $\mathbf{x}(0)$ and (perhaps) some clamps or bias signals. The independent, asynchronous operation of the individual neurons causes a global energy function to decrease until convergence to a stable state at the bottom of an "energy valley." The result is what Cohen and Grossberg (1983) call global pattern formation.

Computational Energy

Consider the quadratic form:

$$H(x) = -(½)\, x^T W x - z^T x \ , \qquad \textbf{(2a)}$$

in which **x** is the Nx1 network state vector, **W** is the NxN weight matrix, **z** is the Nx1 bias vector, and the superscript, **T**, denotes transposition. Written in terms of the components:

$$H(\mathbf{x}) = -(½)\sum_{i=1}^{N}\sum_{j=1}^{N} W_{ij}x_i x_j - \sum_{i=1}^{N} z_i x_i$$

(2b)

is the same as equation (9) of the previous chapter, except that the symbol H is substituted for symbol E. Since the physics literature is consistent in using H for the Hamiltonian (which is usually equal to the energy in classical mechanics), and because we shall need the operator E to denote expected values, this notation might be preferable — at least until we need a symbol for the information-theoretic entropy of the network state! Observing that H(**x**) is the Lyapunov function of a symmetrically interconnected network of binary threshold units subject to the continuous dynamics of the Cohen-Grossberg system, the evolution of **x**(t) will proceed from higher to lower values of H(**x**), ending in stable states at local minima.

Yet it is not necessary to study the stability of differential equations in order to derive or understand this result. Hopfield (1982) stated and proved the principle in two sentences with two equations. That proof, however, is perhaps too compact, leaving some important points understated. Hopfield's argument can be elaborated as follows.

Theorem 3.1: The computational energy function, H(**x**), of a network of McCulloch-Pitts neurons will be decreasing under the dynamics if all of the following conditions are satisfied:

(a) reciprocal connections. The weight matrix is symmetric ($W_{ij} = W_{ji}$);

(b) the absence of self-stimulating connections. $W_{ii} = 0$ for each i;

(c) asynchronous dynamics. No more than one neuron can change state at any given time.

Proof: Let $\mathbf{x}$ and $\mathbf{x}'$ be two state vectors with corresponding energies $H = H(\mathbf{x})$ and $H' = H(\mathbf{x}')$. Then the change in H that occurs in going from $\mathbf{x}$ to $\mathbf{x}'$ is:

$$\Delta H = H' - H$$

$$= -(\tfrac{1}{2})\sum_i \sum_j W_{ij}(x_i' x_j' - x_i x_j) - \sum_i z_i (x_i' - x_i) \quad .$$

By the symmetry of $\mathbf{W}$:

$$\sum_i W_{ij} x_i = \sum_i W_{ji} x_i = y_j \quad ,$$

the second equality being the definition of the net input according to equation (1) of chapter 1. Then:

$$\Delta H = -(\tfrac{1}{2})[\sum_j y_j \, \Delta x_j + \sum_i y_i \, \Delta x_i] - \sum_i z_i \, \Delta x_i + Y$$

$$= -\sum_i u_i \, \Delta x_i + Y \quad , \qquad \textbf{(3a)}$$

where $u = y + z$ is the total input and:

$$Y = -(\tfrac{1}{2})\sum_i \sum_j W_{ij} \, (\Delta x_i)(\Delta x_j) \quad . \qquad \textbf{(3b)}$$

Suppose that *only one unit* can change at a time. Then $W_{ii}=0$ implies $Y = 0$ and the energy change is:

$$\Delta H = -u_i \Delta x_i = -u_i [\mathcal{H}(u_i) - x_i] \qquad \textbf{(4)}$$

when i is the unit in question. If $\Delta x_i = 0$, then $\Delta H = 0$. If not, then either $x_i = 0$ and $\mathcal{H}(u_i) = 1$ (the neuron turns ON) or $x_i = 1$ and $\mathcal{H}(u_i) = 0$ (the neuron turns OFF). In the first case, the net influence of the other neurons must be positive ($u_i > 0$) so both factors are positive and ΔH is negative. In the second case, $u_i < 0$ and $\mathcal{H}(u_i) - x_i$ is also. Hence, both factors are negative and $\Delta H < 0$.

As long as the units all obey the excitation-activation relation of the McCulloch-Pitts neuron, and update their states one at a time, every change in H that accompanies the toggling of a unit will be zero or negative. QED

To show that each of these conditions is necessary, we show that if any one of them is violated, it becomes possible to find a positive energy change.

Let condition (c) be put aside, so that two units (#1 and #3) can change simultaneously. Then:

$$Y = -W_{13}(\Delta x_1)(\Delta x_3)$$

and

$$\Delta H = -u_1 \Delta x_1 - \Delta u_3 \Delta x_3 + D$$

where the symmetry of W has been used.

For example, let N = 3, $\mathbf{z} = (-\epsilon, 0, -\epsilon)$ and:

$$\boldsymbol{W} = \begin{pmatrix} 0 & +1 & -1 \\ +1 & 0 & +1 \\ -1 & +1 & 0 \end{pmatrix} .$$

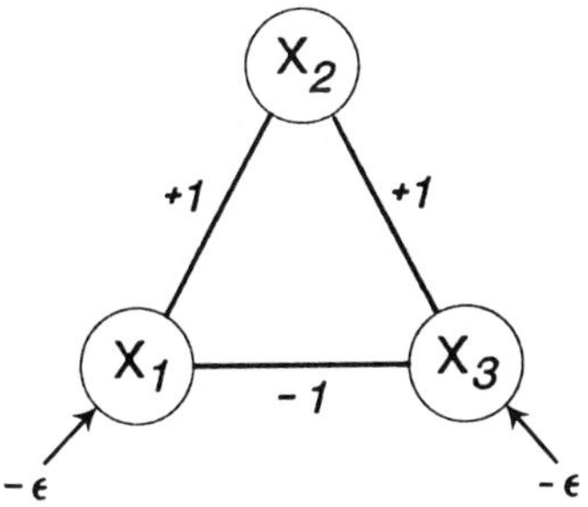

Starting from a state $\mathbf{x} = (1,1,1)$, since $u_1 = y_1 - \epsilon = 0 - \epsilon = u_3$, units 1 and 3 will both turn OFF. The energy before is:

$$H = -\tfrac{1}{2} \sum_{i=1}^{3} \sum_{j=1}^{3} W_{ij}\, x_i\, x_j - \sum_{i=1}^{3} z_i\, x_i$$

$$= -\tfrac{1}{2} \sum_{i=1}^{3} \sum_{j=1}^{3} W_{ij} - \sum_{i=1}^{3} z_i$$

$$= -1 + 2\epsilon.$$

Afterwards, H = 0. If $\epsilon < \tfrac{1}{2}$ the energy has increased. Synchrony allows H to increase.

Condition (b) of the theorem is also required since only $W_{ii} = 0$ causes D to vanish. If we take:

$$W = \begin{pmatrix} -1 & 3/4 \\ 3/4 & 0 \end{pmatrix} \quad x = \begin{pmatrix} 1 \\ 1 \end{pmatrix}$$

then the network state vector is calculated as:

$$y = \begin{pmatrix} -1 & 3/4 \\ 3/4 & 0 \end{pmatrix} \begin{pmatrix} 1 \\ 1 \end{pmatrix} = \begin{pmatrix} -1/4 \\ 3/4 \end{pmatrix}$$

$$x' = \mathcal{H}(y) = \begin{pmatrix} 0 \\ 1 \end{pmatrix}$$

The energy increases since:

$$H(x') = -\tfrac{1}{2} \sum_{i=1}^{2} \sum_{j=1}^{2} W_{ij}\, x_i\, x_j$$

$$= -\tfrac{1}{2}\, W_{22}\, x_2\, x_2 = 0$$

and:

$$H(x) = -\tfrac{1}{2} \sum_{i=1}^{2} \sum_{j=1}^{2} W_{ij}\, x_i\, x_j$$

$$= -\tfrac{1}{2} \sum_{i=1}^{2} \sum_{j=1}^{2} W_{ij}$$

$$= -\tfrac{1}{2}\,(-1 + 3/2) = -\tfrac{1}{2}\,(\tfrac{1}{2}) = -\tfrac{1}{4} < H(x') \,.$$

If condition (a) alone is violated, the change in energy takes the problematic form:

$$\Delta H = -½ \, (u_i + z_i + \sum_j W_{ji} x_j) \, \Delta x_i \quad .$$

Later in this chapter we will examine some asymmetric networks that fail to converge and do not minimize energy. So we postpone further discussion of this point and simply observe that: all three conditions of the theorem are needed to guarantee energy minimization.

Problem 3.1 : Show that symmetric networks of *sign neurons* also minimize computational energy.

Solution: Equation (4) becomes $H = -u_i x_i = -u_i[\mathrm{sgn}(u_i) - x_i]$. If **x** and **x**′ are *not* the same, this is $H = -2u_i \, \mathrm{sgn}(u_i) = -2|u_i| \leq 0$.

The computational energy, with reference to equations (2), has a quadratic part, $-(1/2)\mathbf{x}^T\mathbf{W}\mathbf{x}$, and a linear part, $-\mathbf{z}^T\mathbf{x}$.

Problem 3.2 : Show that, for a network of McCulloch-Pitts neurons, the quadratic part of the computational energy is the sum of all the weights on links between *pairs* of active units.

Solution: The quadratic part is:

$$-(½)\sum_{j=1}^{N}\sum_{i=1}^{N} W_{ij}x_ix_j = -\sum_{j=i+1}^{N}\sum_{i=1}^{N-1} W_{ij}x_ix_j = -\sum_{i<j} W_{ij}x_ix_j$$

by the symmetry of **W** and the fact that the diagonal elements are all zero, where the second equality serves to define the often used notation. That is to say: a sum over all "i<j" is a sum over all distinct (unordered) *pairs* of units.

Illustration: Consider the network of seven McCulloch-Pitts neurons depicted in fig. 3.1. Starting from an arbitrary initial state, pick a unit at random and update its state. For instance, initial state (1111111), with energy H=-5, can lead in two steps to the stable state (0111101), with energy H=-7, s units #1 and #6 turn OFF consecutively.

Problem 3.3: Let a single pattern $\mathbf{x}^o = \mathbf{v}$ be stored by means of the outer product rule in a network of McCulloch-Pitts neurons. Show that **v** is stable in the absence of any biases (or thresholds). Assume that the activity of this state Chapter 1, equation 7) is at least two.

Solution: Stability is assured when H(**v**) is a local minimum. Since:

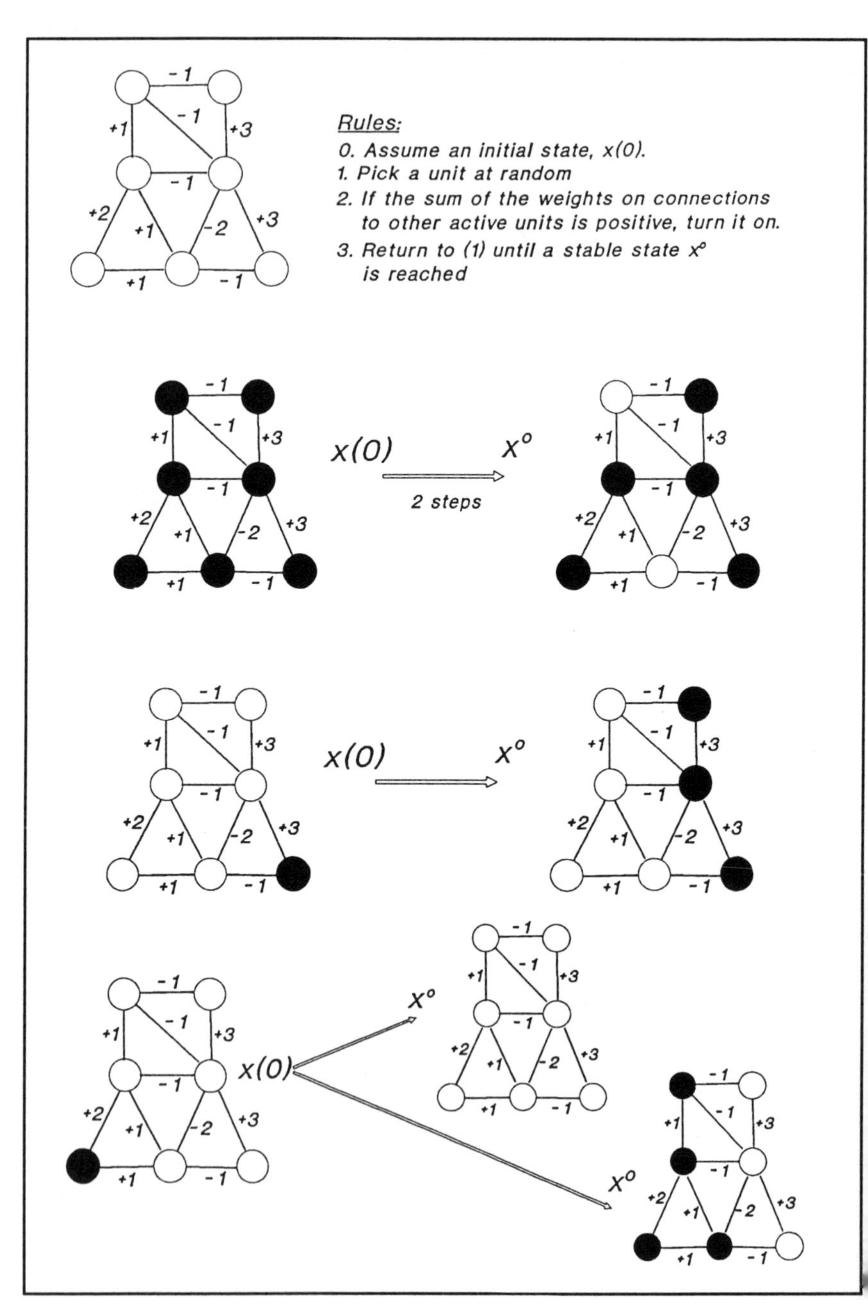
Rules:
0. Assume an initial state, x(0).
1. Pick a unit at random
2. If the sum of the weights on connections to other active units is positive, turn it on.
3. Return to (1) until a stable state x° is reached
x(0)
x°
2 steps
-1
+1
+3
+2
-2

Figure 3.1 A simple Hopfield net. Units have binary states. Connections are symmetric.

$$H(x) = -(½)\sum_i \sum_j (2v_i-1)(2v_j-1)x_i x_j$$

$$= -(½)[\ \sum_i (2v_i-1)x_i\ ]^2$$

when the outer product rule is used. If $\mathbf{x} = \mathbf{v}$, then, since $v_i \in \{0,1\}$, $v_i^2 = v_i$ and:

$$\sum_i (2v_i-1)v_i = \sum_i v_i = A \geq 2$$

for $v_i \in \{0,1\}$, the energy of the stored state is $H(v) = -A^2/2$. If $\mathbf{x}$ differs from v in a *single component*, p, then:

$$\sum_i (2v_i-1)x_i = \sum_{i \neq p} (2v_i-1)v_i + (2v_p-1)x_p$$

$$= A - (2v_p-1)v_p + (2v_p - 1)x_p$$

$$= A + (2v_p-1)(x_p-v_p) = A-1$$

and the *energy gap* is:

$$H(x) - H(v) = -(A-1)^2/2 + A^2/2 > 0 \quad .$$

Thus state $v = \mathbf{x}^o$ is indeed an energy minimum.

Problem 3.4: Consider the network of four McCulloch-Pitts units shown here, in which units #3 and #4 have indicated biases, and units #1 and #2 are clamped by inputs of z_1 and z_2, respectively. More specifically, let z_1 (or z_2) be +10 to clamp a unit ON, and -10 to clamp a unit OFF. For each of the four possible clamped conditions (x_1, x_2) that minimizes the computational energy, construct a "truth table" x_3 (x_1, x_2) based on the result.

Solution: The table on the left below gives the energy of each substate for each of three input conditions. (Since the weights are not altered by interchanging units #1 and #2, case (x_1, x_2) = (1,0) is the same as (0,1).) For each possible value of (x_1, x_2), search the column for the smallest energy. For (0,0), this is 0. For (0,1) it is -11.5 and for (1,1), -25. Since the net moves toward minimum energy states, when (x_1, x_2) is (0,0), $x_3 = x_4 = 0$. When (x_1, x_2) is (1,0) (and, by symmetry when it is (0,1)) $x_3 = x_4 = 1$. Finally, for

(1,1), $x_3 = 0$ and $x_4 = 1$. The resulting truth table (right) for x_1, x_2, and x_3 is that of the exclusive-or function ($x_3 = x_1$ .EOR. x_2). The network is the same as the one in the lower right quadrant of fig. 1.4.

x_3	x_4	(x_1, x_2)		
		(0,0)	(0,1)	(1,1)
0	0	0.	-10.0	-20.0
0	1	3.0	-11.0	-25.0
1	0	1.0	-7.5	-16.0
1	1	1.0	-11.5	-24.0

x_1	x_2	x_3
0	0	0
0	1	1
1	0	1
1	1	0

Convergence Rates

In describing the properties of his original stochastic model, Hopfield (1982) observed that convergence to stable states typically required a time of 4/R, where R is the "mean attempt rate" (in updates per neuron per unit time). Hopfield used the symbol W for this mean processing rate; but here we are using **W** for the symmetric NxN weight matrix defining the connectivity of N neuron-like units: $W_{ij} = W_{ji}$ and $W_{ii} = 0$ (Hopfield's **T**). These N McCulloch-Pitts neurons in Hopfield's stochastic model updated their states $x_1,\ldots,x_N$ {0,1} asynchronously, according to the rule:

$$x_i = \begin{cases} 1 \\ 0 \end{cases} if \sum_{j=1}^{N} W_{ij} x_j - \theta \begin{cases} >0 \\ \leq 0 \end{cases},$$

where the threshold θ was set to zero in most cases. Each neuron invoked this test at random times, independent of all the other neurons, and at the same mean rate R. Clearly the mean number of updates per unit time is NR in this network of N concurrent processing elements. Convergence in 4/R units time then means that about 4N updates occur in going from the (random) initial state to the (final) stable state: (N units) x (R updates/neuron/sec) x (4/R sec) = 4N updates.

Another way to express the dynamics of the stochastic model is to count time in discrete steps (t=0,1,2,...) and to write the iterated equation:

$$x_i(t+1) = \begin{cases} \mathcal{H}\left[\sum_{j=1}^{N} W_{ij}\, x_j(t) - \theta_i + z_i\right] & \textit{if } i = n(t) \\ x_i(t) & \textit{otherwise} \end{cases} \qquad (5)$$

where {n(t), t=1,2,3,...} is a sequence of independent random numbers, (uniformly distributed among the first N integers) that index the neurons. (Recall Fig. 2.2.) To see this, assume that updates are instantaneous events, separated in time by intervals of random duration. Let T(i,k) be the time (continuously distributed between the times zero and T_{max}) at which unit i experiences its kth update. Then the elements of the array:

T(1,1) T(1,2) T(1,3) ...
T(2,1) T(2,2) T(2,3) ...
: : :
T(N,1) T(N,2) T(N,3) ...

(which are approximately NRT_{max} in number) can be placed in ascending order:

$$0 < T(i_1,k_1) < T(i_2,k_2) < \ldots .$$

The sequence i_1, i_2, i_3,... then consists of independent random numbers that are identically and uniformly distributed on {1,...,N}. (Proof: Independence is given; and the uniform distribution of the indices is a necessary consequence of all neurons having the same mean processing rate.)

Thus a convergence time of 4/R, corresponding to 4N updates, also corresponds to 4N iterations of equation (5), which can be programmed rather directly on a general purpose computer.

Definition: In a network of N neurons, N consecutive iterations of equation (5) constitute a (random) *sweep* of the network. Then Hopfield's casual observation can be paraphrased by saying that, in symmetric networks (with N ≈ 30), the convergence time is typically four sweeps. Hopfield and Tank (1986c, p. 233) later suggested that convergence time (in sweeps) might be the same irrespective of the number of neurons.

Experiment: To test this hypothesis in numerical experiments, it will be necessary to define the conditions more precisely. In general the procedure will be as follows:

(a) Create a symmetric matrix **W** (NxN) with all zeros on the diagonal.
(b) Create an **x**(0) of N independent binary components that are 0 or 1 equiprobably.
(c) Iterate equation (1) until a stable state is reached, or a "suitably large" number of steps t_{max} has failed to find one. Record the number of time steps and
(d) repeat these procedural steps a large number of times for each N.

The result of this procedure at a particular N can be expressed as an accumulated histogram (or sample distribution function) showing the number (or fraction) of networks converged at each time step. Fig. 3.2 exemplifies this with N=10. The ascending staircase that lies to the left in the figure summarizes 100 trials each using a different symmetric weight matrix in which the upper triangle consisted of independent, zero-mean, logistic random variables. The fitted curve is a log-logistic distribution function. Since the 50% convergence level is reached just after t=20 iterations, the median convergence time, denoted t_{50}, is slightly more than 20 iterations, or 20/N = 2 sweeps. The staircase which lies to the right in fig. 3.2 was obtained by a similar procedure, except that the symmetry of **W** was revoked, and all the nondiagonal elements were independent and identically distributed. The asymmetric networks seem to take much longer to converge than symmetric ones!

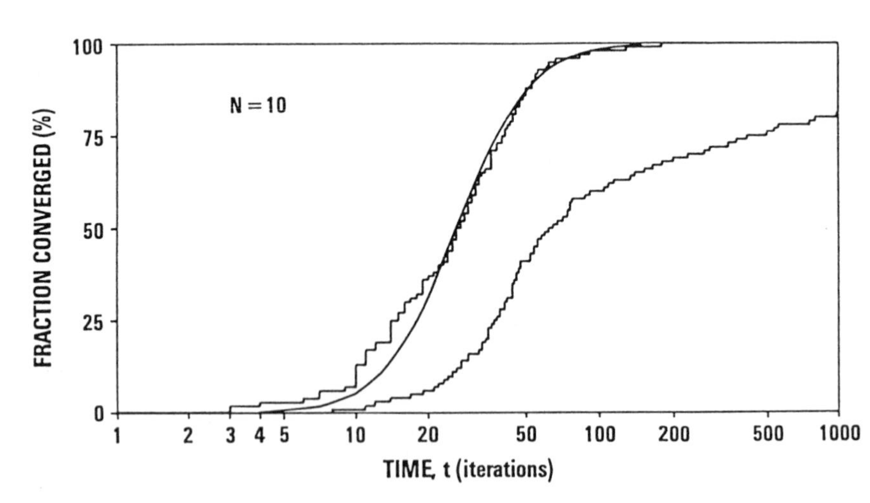

Figure 3.2. Fraction converged versus time (steps) for networks of ten neurons, symmetric (above) left) and asymmetric (right).

Random Symmetric Weights

The open circles plotted in fig. 3.3 show the median sweeps to convergence for a range of N, using the same stochastic rule as just noted for generating the symmetric **W** in step (a) of the procedure. With a sample size of more than 100 independent trials per plotted point, the increasing trend in $t_{50}(N)$ is statistically significant. (For N=33, the result is closer to five sweeps than to four.) The fitted curve is:

$$t_{50}/N = [(N/6)\ln(2N)]^{1/2} \text{ sweeps.}$$

Random Stored Patterns

The squares plotted in fig. 3.3 show the median sweeps to convergence for several values of N using a different rule for generating symmetric weight matrices. In particular, let $\mathbf{x}^1,\ldots,\mathbf{x}^M$ be vectors of N independent bits that are equiprobably zero or one; and use the outer product rule for the off-diagonal components of **W**. This might complicate the experiment, since step (d) of the procedure (see p. 62) needs to be "for each N and M." The plotted points are for M=N/10, i.e., for (M,N) = (1,10), (2,20), (3,30)... It remains to be seen whether the median convergence time levels off around four sweeps for large N in this case.

The triangles plotted in fig. 3.3 above the abscissa N=20 are for the indicated values of M. As the number M of patterns stored (by the outer product rule) increases, the median convergence time likewise grows, apparently exceeding the result for random, symmetric **W**.

These results are in conflict with the findings of Crisanti and Sompolinsky (1988), whose Cray X-MP simulations with N ranging from 100 to 1000 led them to very different conclusions. In step (a) of the procedure, they used off-diagonal weights:

$$W_{ij} = S_{ij} + \kappa A_{ij} ,$$

with S_{ij} and A_{ij} being independent, zero-mean, normal random variables, where $S_{ji} = S_{ij}$ is the symmetric part and $A_{ji} \veebar A_{ij}$ is the asymmetric part. (Symbol $\veebar$ denotes statistical independence.) The parameter $\kappa \in [0,1]$ is the degree of asymmetry. Crisanti and Sompolinsky (p. 37) observed that the average logarithm of the convergence time grows with the degree of asymmetry as $Na(\kappa)$, where $a(\kappa) \to 0$ as $\kappa \to 0$ and $a(\kappa) \to \infty$ as $\kappa \to 1$. In other words, for random, symmetric weights, Hopfield's remarks, to the effect that convergence time is the same for all N, were thoroughly affirmed! The fact that Crisanti and Sompolinsky used sign neurons instead of McCulloch-Pitts units is probably irrelevant to this conclusion.

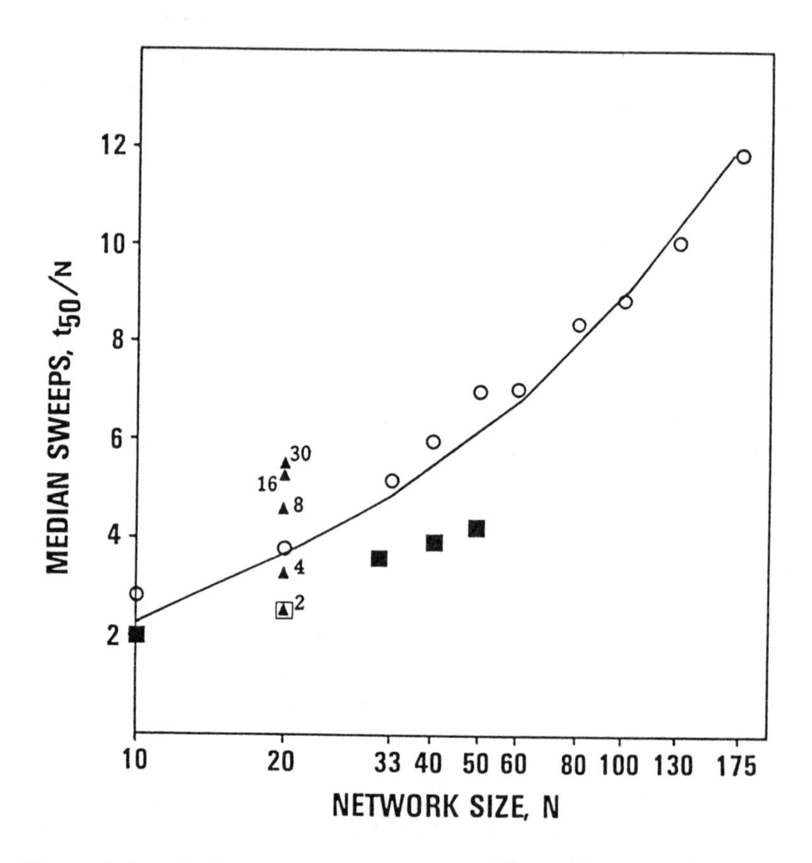

Figure 3.3. Median convergence time (in sweeps) versus network size for indicated numbers of stored patterns.

Problem 3.5: The Blindsider Mk 4 air-launched antitank missile uses a Hyperfast Series III on-board computer for automatic target recognition. The Series III is a serial machine that executes 100 Gigaflops (10^{11} floating point operations per second). It simulates a Hopfield net (using the asynchronous dynamics) in which there is one binary threshold unit for each pixel in a 100-by-1000 scanned image.

(a) If the time to convergence is 4.0 sweeps of the net, how many seconds does convergence take in the on-board computer?
(b) What if the convergence time is $[(N/6)\ln(2N)]^{1/2}$ sweeps?

Solution: The 100 x 1000 $= 10^5$ pixels are one-to-one with the $N = 10^5$ eurons. Each iteration of equation (5) requires $N - 1 \approx N$ flops (floating point perations) of the form $y_i \leftarrow y_i + W_{ij}x_j$. Thus each sweep takes N^2 flops. ssuming the execution of all the other operations is instantaneous by omparison, the time per sweep is:

$$[(10^5)^2 \text{ flops}]/[10^{11} \text{ flops/sec}] = 0.1 \text{ sec.}$$

hen the convergence time is 0.4 seconds in case (a). On the other hand, if onvergence time scales with N as suggested, the median number of sweeps to onvergence is about 451, corresponding to 45.1 seconds in case (b).

The Markov Property

The dynamical equation (5) produces a sequence $\{\mathbf{x}(t), t=0,1,2,\ldots\}$ of ates that has the Markov property: For any $\mathbf{x}$ in the state space of the network, e probability of having $\mathbf{x}(t+1) = \mathbf{x}$, given the entire past history $\mathbf{x}(0),\ldots,\mathbf{x}(t)$ the process, is the same as the probability conditioned on $\mathbf{x}(t)$ alone:

$$\Pr[\mathbf{x}(t+1) = \mathbf{x} \mid \mathbf{x}(t), \mathbf{x}(t-1),\ldots,\mathbf{x}(0)] = \Pr[\mathbf{x}(t+1) = \mathbf{x} \mid \mathbf{x}(t)].$$

ne evolution of the network state is determined, at each step, by the present ate, the weights and inputs, and the random number generator that picks the dex n(t) of the next unit updated.

The (Hamming) *distance* between two states in a network of binary nits is the number of components (or units) in which they differ. See p. ?.

Equation (5) constrains the evolution of the network state to proceed no ster than one per step (since it permits no one-step transitions of distance >1). ppendix A shows how the one-step transition matrix $\mathbf{P}$ of the Markov chain (t), $t=0,1,2,\ldots\}$ can be written explicitly in terms of the weights and states. gure 3.4 shows the nondiagonal elements of $\mathbf{P}$ in the simplest case, $N=2$ and $=0$ (no biases). The Appendix recalls that the states of a Markov chain are ther *transient* or *persistent*. For transient states:

$$\Pr[\mathbf{x}(t)=\mathbf{x} \mid \mathbf{x}(0)] \rightarrow 0 \text{ irrespective of } \mathbf{x}(0).$$

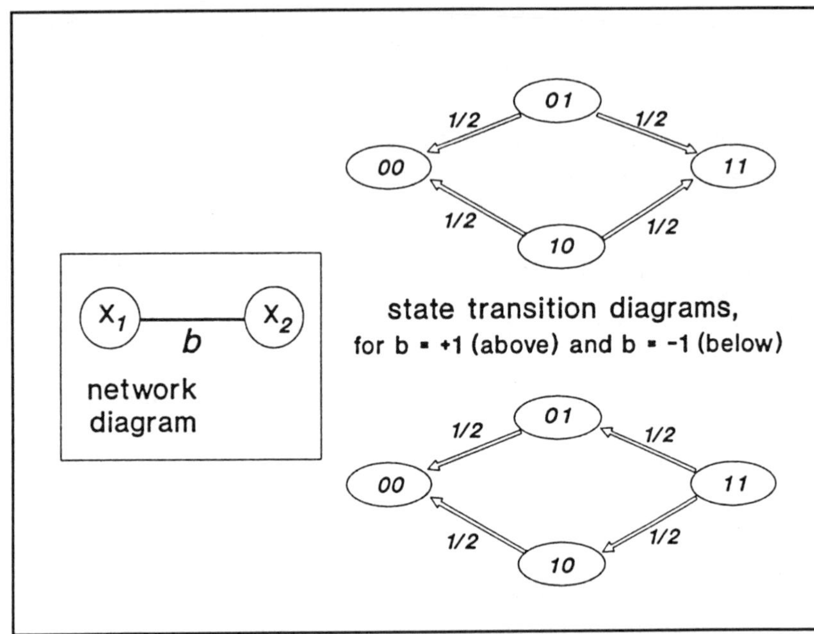

Figure 3.4 Two-unit network and state transition diagrams.

For persistent states, the limiting probability may be zero or positi depending on the initial state; but, for any $\mathbf{x}(0)$ whatever, the state $\mathbf{x}(t)$ ultimately *absorbed* into ("attracted" by) an *irreducible (closed) set of persiste states*. (See, for instance, the gambler's ruin problem where bankruptcy is persistent state.) Once the state enters an irreducible, persistent set, it can nev escape. When the cardinality of such a closed set (C) is unity, the sing absorbing state in question is *stable*. When the cardinality is greater than unit $\mathbf{x}(t)$ can revisit each of the elements of C ad infinitum.

Theorem 3.2: When $\mathbf{W}$ is a (real) symmetric matrix with all zeros the diagonal, all the irreducible, persistent sets of states of the Markov cha $\{\mathbf{x}(t), t=0,1,2,\ldots\}$ induced by equation (5) have cardinality one.

Proof: The conditions on $\mathbf{W}$ are the same as those in theorem 3.1 p. 53. Thus the Markov chain moves downhill in energy until a stable state reached. These stable states collectively exhaust the persistent states of t Markov chain. QED.

When **W** is asymmetric, the persistent states of the Markov chain {**x**(t)} may belong to closed sets of cardinality greater than one. We shall not attempt to prove this, but content ourselves with an example. Refer back to fig. 1.3 of Chapter 1. Let a = -1, b = 2, c = 4, d = 1, ϵ = 0, and δ = -0.01. Assume McCulloch-Pitts neurons and let **x**(0)=(1,1,1). Then **x**(t) returns to the initial state in the following manner:

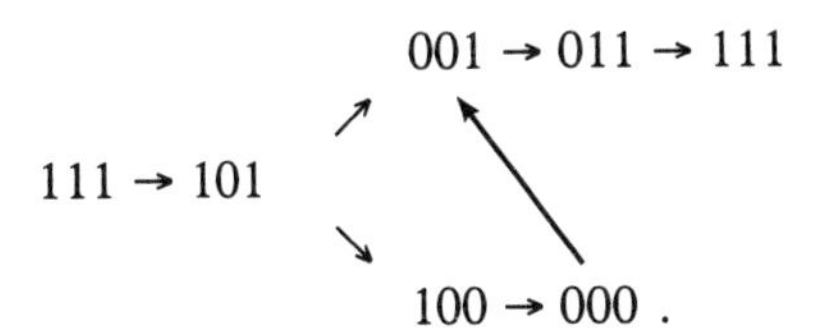

For instance, if **x** = (1,0,1), the total inputs to units #1 and #3 are 0 and -0.9, respectively. Regardless which unit (#1 or #3) is selected for turnoff by the random number generator, the state eventually reaches (0,0,1), which leads back to (1,1,1) as shown. Thus the set {111, 101, 100, 000, 001, 011} is closed. There are no stable states in this network with given bias since the remaining two elements of the state space, 010 and 110, are both transient. Another example, using N = 4 with no biases will be found in appendix A.

Problem 3.6: (Waiting time to absorption): Let the function:

$$index\,[x] = \sum_{i=1}^{N} x_i 2^{i-1}$$

assign a cardinal number to each of the 2^N states of an N-unit McCulloch-Pitts network. Starting from a transient **x**(0), let T^* be the time, in steps of equation (5), when **x**(t) first passes into a persistent state. What is the mean first passage time:

$$e_k = E[T^* \mid \text{index}[\mathbf{x}(0)] = k] \;?$$

Solution: Let P_{km} be the probability of transition in one step from the kth state to the mth state. Let $f_k = e_k + 1$. Refer to Feller (1957), who asks us to prove that:

$$f_m - \sum_{k \in Y} P_{mk} f_k = 1 \qquad \textbf{(6)}$$

for every $\boldsymbol{m} \in \mathbf{Y}$ = {transient states}.

Illustration: Two McCulloch-Pitts neurons are linked by $W_{12} = W_{21} = +1$. In the absence of biases, the stable states are (1,1) and (0,0). With reference to fig. 3.4, the transition matrix is:

$$\boldsymbol{P} = \begin{bmatrix} 1 & 0 & 0 & 0 \\ \tfrac{1}{2} & 0 & 0 & \tfrac{1}{2} \\ \tfrac{1}{2} & 0 & 0 & \tfrac{1}{2} \\ 0 & 0 & 0 & 1 \end{bmatrix}.$$

Feller's linear system (6) reduces to:

$$f_1 - f_2/2 = 1 \quad \text{and} \quad f_2 - f_1/2 = 1 \ ,$$

which is solved by $f_1 = f_2 = 2$; so the mean convergence time is $e = f\text{-}1 = 1$ for transient states (0,1) and (1,0). The variance is zero, since one step of equation (5) produces (0,0) — if n(0) is the index of the ON unit — or (1,1).

Problem 3.7: Consider a pair of McCulloch-Pitts neurons networked by an antisymmetric weight matrix with W_{12} =-1 and W_{21} = 1. What states are stable? Calculate the mean time to absorption from each transient state.

Solution: The transition matrix on the right shows that only (0,0) is stable. Feller's equations (6) are solved for the three unknowns f_1=3, f_2=2, and f_3=4.

$$\begin{bmatrix} 1 & 0 & 0 & 0 \\ \tfrac{1}{2} & 0 & 0 & \tfrac{1}{2} \\ \tfrac{1}{2} & 0 & \tfrac{1}{2} & 0 \\ 0 & 0 & \tfrac{1}{2} & \tfrac{1}{2} \end{bmatrix}$$

Continuation: What is the mean time to convergence in the sense defined earlier?

Answer: The mean number of steps to convergence from each possible initial state is listed thus: e_0=0, since state (0,0) is stable; $e_1 = f_1\text{-}1 = 2$; e_2 =

; and $e_3 = 3$. If the initial state is equiprobably any one of these, the mean onvergence time is:

$$(1/4)\sum_{m=0}^{3} e_m = 3/2 \textit{ iterations}$$

r 0.75 sweeps.

Basins of Attraction

The energy valley surrounding a stable state $\mathbf{x}^o$ in a symmetrically nterconnected (collective computation) net might be impossible to envision. Yet e may discuss it in terms of basins of attraction that are characterized by a vertical" depth and a "longitudinal" extent. The depth is fairly easy to easure, since energy defines the vertical scale. Longitudinal extent is more roblematic, though its motivation is quite clear. Fig. 3.5 depicts the energy ndscape in terms of (dashed) contours — lines of constant H — and (solid) ajectories that terminate at stable states. Hopfield and Tank (1986c) invoked is kind of visualization under the assumption of the continuous dynamics of the ectronic network and the nonlinear differential equations describing it. ecause this process is asynchronous, the motion is strictly downhill in energy. ecause it is deterministic, every $\mathbf{x}(0)$ leads to one and only one $\mathbf{x}^o$. (Although earby $\mathbf{x}(0)$ may lead to very different $\mathbf{x}^0$.) The continuous phase space is artitioned into distinct regions belonging to particular stable states. Assuming nary units, the size of the basin of $\mathbf{x}^o$ could be measured as:

$$\#\{\mathbf{x}(0): x(t) \rightarrow \mathbf{x}^o \text{ as } t \rightarrow \infty\}/2^N , \qquad (*)$$

e fraction of the total state space belonging to $\mathbf{x}^o$.

Komlos and Paturi (1988) define the *radius of attraction* of $\mathbf{x}^o$ as the rgest value of r such that every $\mathbf{x}$ at (Hamming) distance not more than rN om $\mathbf{x}^o$ eventually reaches $\mathbf{x}^o$. Assuming that M patterns are stored with the ter product rule, they established the existence of a positive radius about the ndamental memories when $M < N/(4\ln N)$ and showed that both synchronous d asynchronous dynamics give convergence (which takes $O(\ln(\ln N))$ nchronous sweeps).

Also with reference to the storage of M patterns using the outer product le, Hopfield (1982) had observed that "memories too close to each other are nfused and tend to merge.... The case $N = 100$, $M = 8$, was studied with

7 random memories and the eighth made up a Hamming distance of only 30, 2 or 10 from one of the other seven. At a distance of 30, both similar memori were usually stable. At a distance of 20, the minima were usually distinct b displaced. At a distance of 10, the minima were often fused." The Komlo Paturi radius of attraction is a refinement of this semiquantitative descriptio They showed that, for some constants r and a, if $M \leq aN$ and if $\mathbf{x}(0)$ is with distance rN from a stored $\mathbf{x}^o$, then, in about $\ln(N/M)$ synchronous sweeps, t evolution of $\mathbf{x}(t)$ will terminate within distance $Ne^{-N/4M}$ of $\mathbf{x}^o$. Moreover, wh $M < N/(4\ln N)$, the bounds on the radius of attraction are $0.024 < r < .12$

Problem 3.8: A data source produces sequences of symbols from twenty-six letter alphabet. Each letter is transmitted separately as a 1000-b block of code. At the output of a noisy channel, the decoder is a Hopfield n in which the twenty-six letter codes are the stored patterns. Assuming a > 0.026, how many bit errors per block can be corrected by this network?

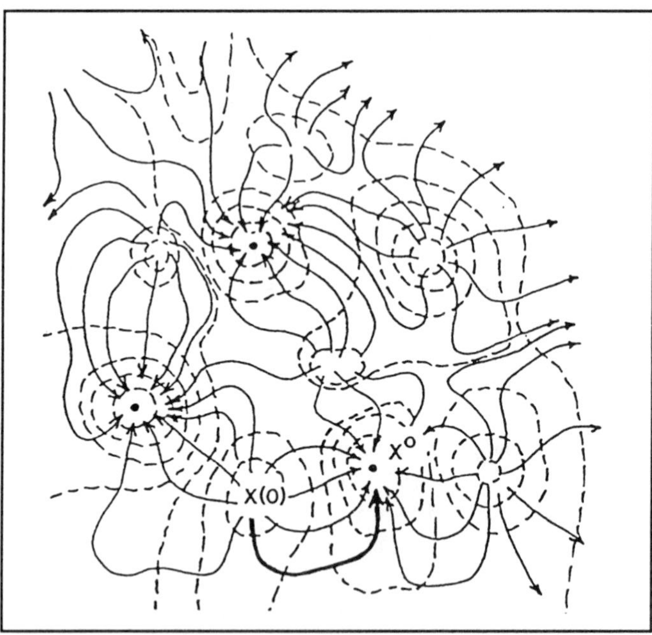

Energy landscape dominated by stable states and their basins of attraction. Network stat evolves from $\mathbf{x}(0)$ to $\mathbf{x}^0$ going downhill in energy

Figure 3.5 Stable states and their basins of attraction dominate the energy landscape.

Solution: With (N,M) = (1000,26), the stable state to which the network is attracted lies within $Ne^{-N/4M} \approx 0$ bits from the truth. Since $N/(4\ln N) = 36 > 26$ (= M), at least 0.024N = 24 bits can be corrected. In any case, a bit error rate of more than N/8 = 125 per block will probably defeat the decoder.

Another way to measure the basins of attraction is suggested by the treatment of the stochastic model in terms of Markov process theory. The expression (*) of Komlos and Paturi has no meaning in the context of the stochastic model, since a given initial state may lead to different stable states depending on the interrogation sequence {n(t)} in the scheme of equation (5). Feller (1957, sec. XV.8) shows that the probability of eventual absorption by a persistent state (m), starting from a transient state (k) is obtained by solving the system of linear equations:

$$q_k - \sum_{\nu \in Y} P_{k\nu} q_\nu = P_{km} ,$$

where Y (as before) is the set of all transient states and **P** is the transition matrix of the Markov chain. As **P** is $2^N x 2^N$, this system of equations will not be solved by most computers in a contemporary lifetime for N much greater than 30. In principle, however, it serves to define:

$$q_k(\mathbf{x}^o) = \Pr\{\mathbf{x}(\infty)=\mathbf{x}^o \mid \text{index}[\mathbf{x}(0)] = k\}.$$

Let:

$$Q_{mk} = \Pr\{\text{index}[\mathbf{x}(\infty)]=m \mid \text{index}[\mathbf{x}(0)]=k\},$$

where the index is $\Sigma_i\, x_i\, 2^i$ as before. This Q_{mk} is the same thing as $q_k(\mathbf{x}^o)$ when $m = \text{index}(\mathbf{x}^o)$. Let us refer to the states by their indices. A great number of independent trials starting from state k will lead to state m some fraction of the time; and this fraction approaches Q_{mk} as the number of trials goes to infinity (fig. 3.6). The probability of eventual absorption by m, given an initial state that is equiprobably any one of the 2^N possible states, is:

$$Q_m = 2^{-N} \sum_{k=0}^{2^N-1} Q_{mk} . \tag{7}$$

The set $\{Q_k: k=1,2,\ldots,\#\{\mathbf{x}^o\}\}$, which contains one element for each stable state, defines a discrete probability density. Q_m is the stochastic counterpart of the quotient in equation * (p. 68) which describes the partition of the whole state space into domains of attraction belonging to stable states.

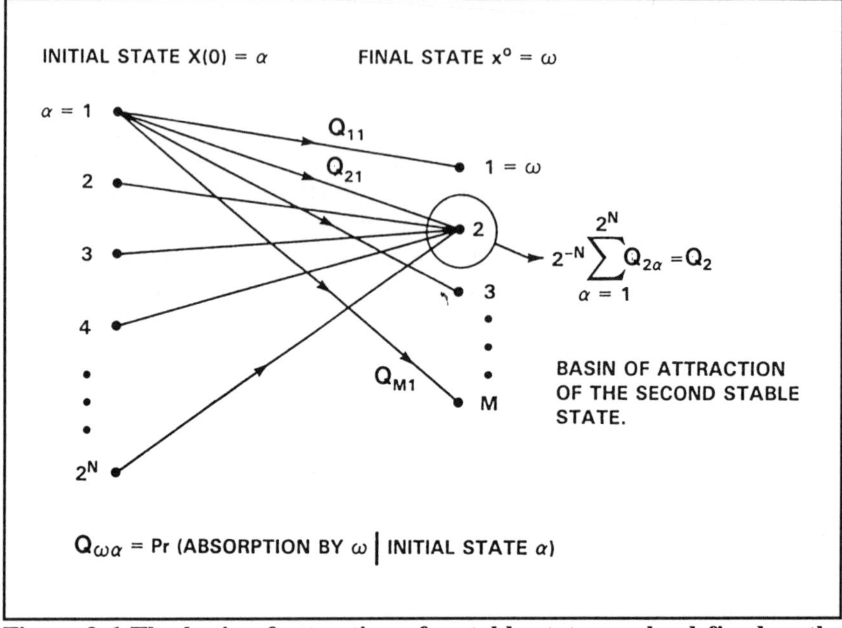

Figure 3.6 The basin of attraction of a stable state can be defined as the probability of eventual absorption into that state, given an initial state that is equiprobably any one of the 2^N possible states.

Problem 3.9: Let there be N=3 McCulloch-Pitts neurons with no biases and:

$$\mathbf{W} = \begin{pmatrix} 0 & 1 & -1 \\ 1 & 0 & 1 \\ -1 & 1 & 0 \end{pmatrix} .$$

What states are stable? Use equation (7) to measure their basins of attraction.

Solution: See fig. 3.7, in which the state transition diagram shows only the nondiagonal elements of **P**. The stable states (from which no arrows emanate) are 000, 110, and 011. All three basins of attraction have the same size, Q = 1/3. To see this, calculate Q_0 and observe that the nonzero stable states must have basins the same size by the symmetry of the network diagram (i.e., its invariance with respect to interchanging indices 1 and 3). With reference to fig. 3.7:

$$Q_0 = (1 + Q_{01} + Q_{02} + Q_{04} + Q_{05})/2^N$$
$$= (1 + 1/2 + 1/3 + 1/2 + Q_{05})/8$$
$$= (14/6 + Q_{05})/8 .$$

Moreover:

$$Q_{05} = (1/3)Q_{01} + (1/3)Q_{04} = 1/3.$$

Hence:

$$Q_0 = 16/48 = 1/3.$$

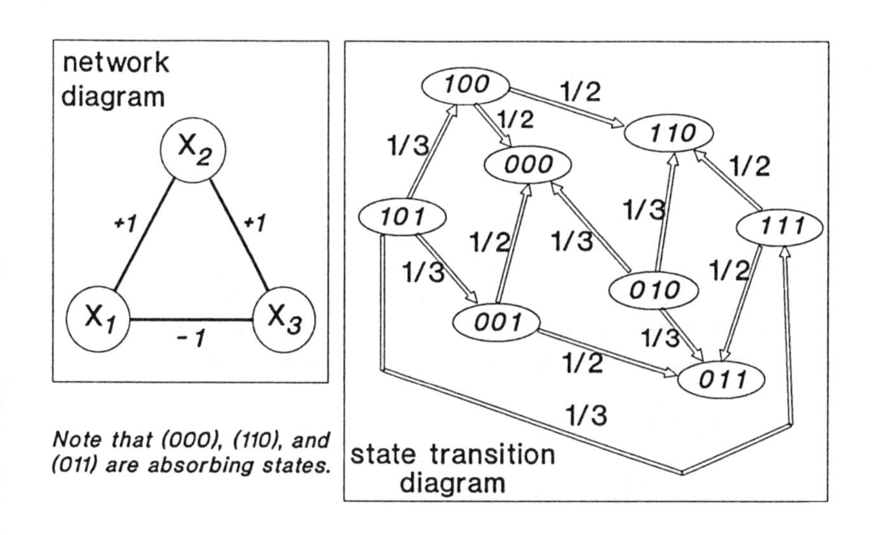

Figure 3.7 Three-unit network and state transition diagram.

Density of Stable States

How many stable states does a typical network have? How are their energy levels distributed? These questions are answered by Nemoto (1988) in an inspired numerical study of randomly generated networks of sign neurons. Let the $\{W_{ij}\}$ in the upper triangle be independent normal random variables with zero mean and variance 1/(N-1), where N (as always) is the number of units. Taking the network state **X** as a random variable, the mean and variance of the energy are:

$$EH(X) = E(-\sum_{i<j} W_{ij}x_i x_j) = -\sum_{i<j} (EW_{ij})E(X_i X_j) = 0$$

and:

$$EH^2 = E[\sum_{i<j} (W_{ij})^2] = [N(N-1)/2]\ Var\ (W_{ij}) = N/2 \quad ,$$

since the expected value of a product $W_{ij}W_{kl}$ is zero. Thus the distribution function (d.f.) of the energy is:

$$Pr(H \leq h) = \Phi(2h^2/N) \tag{8a}$$

in terms of the standard normal d.f.,

$$\Phi(z) = \frac{1}{\sqrt{2\pi}} \int_{-\infty}^{z} \exp(-y^2/2)dy \tag{8b}$$

obtained by invoking a suitable central limit theorem.

The distribution of the energies of *stable* states has the normal form; but it is shifted to the right (uphill!) and more dispersed. Nemoto found that the mean and variance of the energy, conditioned on the stability of the state, are:

$$E[H(\mathbf{X}) \mid \mathbf{X} \in \{\mathbf{x}^o\}] = N/2$$

and:

$$Var[H(\mathbf{X}) \mid \mathbf{X} \in \{\mathbf{x}^o\}] = N.$$

Fig. 3.8 shows the scaled density of stable states for N=24 sign units, the abscissa being $\mathcal{E}$ = (energy - N/2)/ N.

In addition, Nemoto found, for N between 16 and 24 inclusive, that the average number of stable states in a randomly generated (symmetric) net is about:

$$E\ \#\ \{\mathbf{x}^0\} = \exp(0.2N) \quad . \tag{9}$$

For instance, with N=24, the number of stable states is typically 121.

Question: How many stable states should one expect to find in a randomly generated network of N McCulloch-Pitts neurons? (The weights are zero mean, normal random variables.)

THE DENSITY OF STABLE STATES IN A RANDOMLY-GENERATED NETWORK OF SIGN (SPIN) UNITS, N = 24

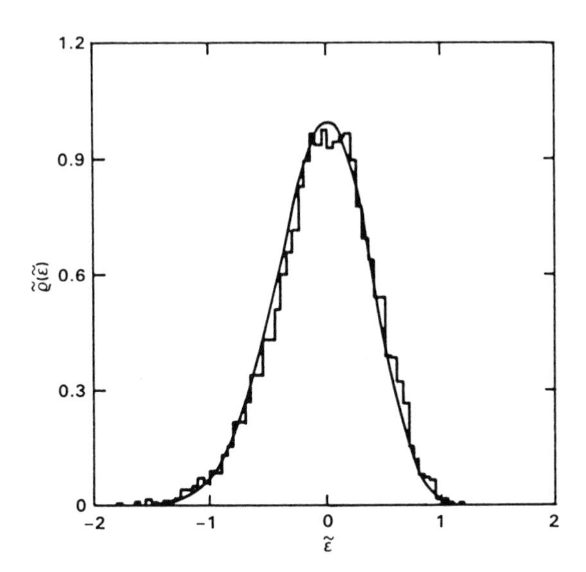

SCALED DENSITY OF METASTABLE STATES $\tilde{\varrho}(\tilde{\varepsilon})$ for $N = 24$, WHERE $\tilde{\varepsilon} = (E - N\varepsilon_o)\, N^{-1/2}$ WITH $\varepsilon_o = 0.506$. THE FULL CURVE REPRESENTS THE ANALYTICAL RESULT OF

$$P_o(q) \cong \left(\frac{\gamma N}{2\pi}\right)^{1/2} \exp -\left(\frac{\gamma N q^2}{2}\right)$$

WITH $\gamma = 1$.

AFTER K. NEMOTO (1988). "METASTABLE STATES OF THE SK SPIN GLASS MODEL," *J. PHYS. A: MATH. GEN. 21*, L287–L294.

Figure 3.8 Scaled density of metastable states.

Best Guess : In the absence of solid theory or hard numerical data, robably about $e^{0.2N}/2$, i.e., half of equation (9). In a net of sign units, -**x** is table if **x** is; but the logical complement of $\mathbf{x} \in \{\mathbf{x}^o\}$ in a McCulloch-Pitts net s not stable in general — unless the weights are obtained by storing random atterns with the outer product rule.

Additional Problems

3.10 Consider this network of six McCulloch-Pitts neurons with symmetric weight matrix:

$$W_{ij} = \begin{cases} -1 & \text{if } |i-j| = 1 \text{ or } 5 \\ +1 & \text{if } |i-j| = 2 \text{ or } 4 \\ 0 & \text{otherwise} \end{cases}$$

(a) Let z = 0.

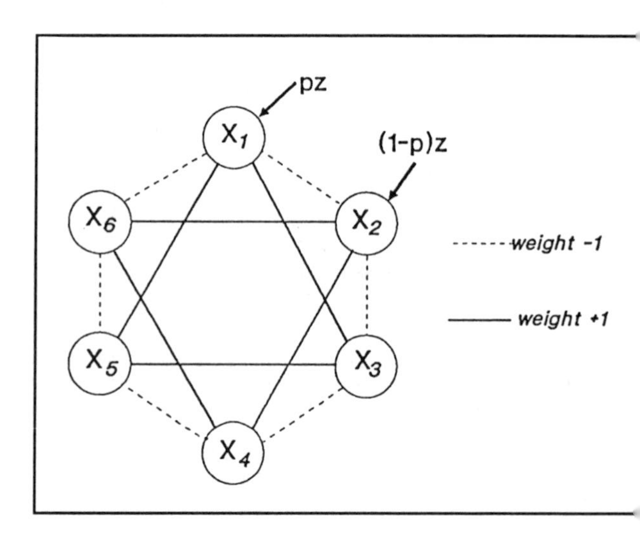

Which states are stable?

(b) Hopfield (1982) says that, given a "confusing stimulus" that is the sum of two vector patterns, a linear associative net will return the sum of the corresponding responses; but a collective computation net will "make a choice" between the two. Let z be nonnegative and assume that $0 < p < 1$. Sketch, in the p, z-plane, the region in which the network will converge to stable state 010101. Assume asynchronous dynamics.

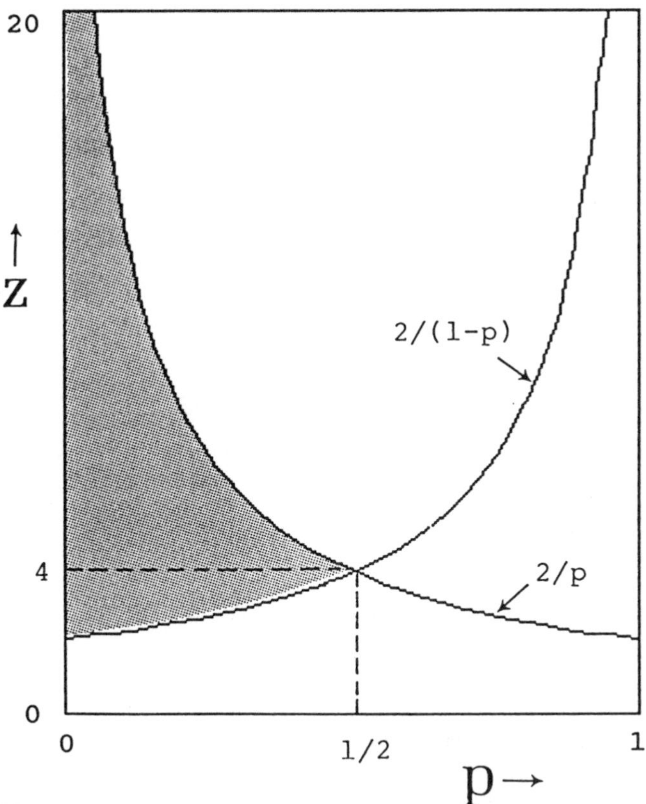

Solution:

(a) The stable states are 000000, 101010, and 010101. Note that, if p = 1/2 and $\theta \approx 0$, equation (3.5) will carry the network from the all-zero state to either one of the nonzero stable states with the same probability (1/2).

(b) A bias greater than 2 will guarantee that a unit remains ON, since the net input cannot be less than -2. The sufficient conditions for evolution to stable state 010101 from initial state 000000 are $(1-p)z > 2$ and $pz \leq 2$. Thus the desired pattern is reached when z lies above the ascending curve 2/(1-p) and below the descending curve 2/p. In the U-shaped region, the network converges to a "confused response," either 110101 or 111010.

3.11

(a) Regarding the network on the left (below), what are the stable states? Which of these attains the global energy minimum?

(b) Answer the same questions for the network below. Assume $C >> 1$.

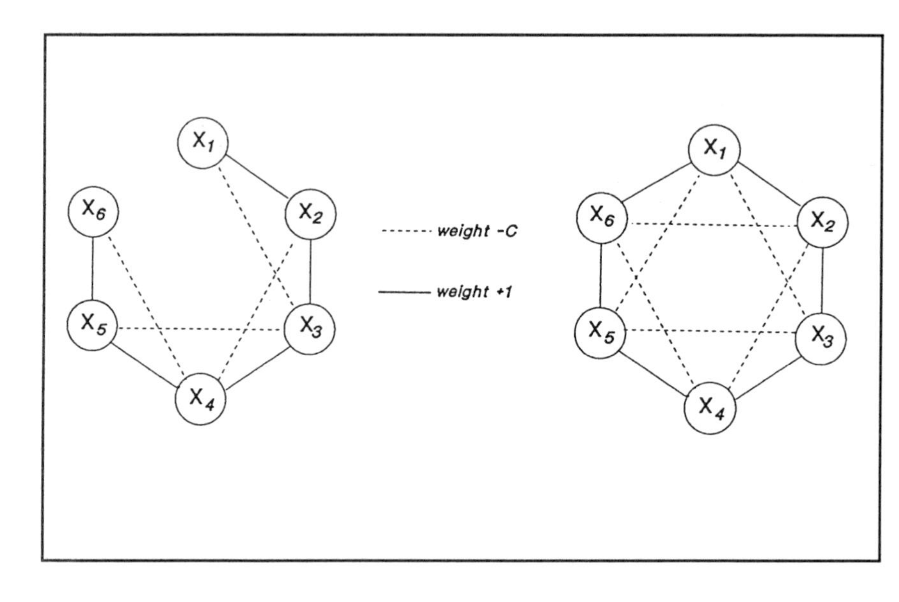

Solution:

(a) The eight stable states include 000000, 110000, and 011000, . . ., 000011, 110011 (the global minimum), and 001100.

(b) When the loop is closed, there are seven stable states; and all the nonzero stable states have the same energy, -1.

3.12 Let there be N McCulloch-Pitts neurons, each pair of which is linked by weights of $-C < 0$. The ith unit has bias $z_i > 0$. Suppose that $\max_{1 \leq i \leq N}(z_i) = A$. The following questions concern the behavior of the network when $A/C << 1$.

(a) Verify the stability of each state of the form:

$$\boldsymbol{x}^{(i)} = (\delta_{1i},\ldots,\delta_{Ni}), \quad 1 \leq i \leq N \quad ,$$

where the Kronecker symbol:

$$\delta_{ij} = \begin{cases} 1 & \textit{if} \;\; i=j \\ 0 & \textit{otherwise} \end{cases}$$

What is the energy of state $\mathbf{x}^{(i)}$?

(b) Starting from $\mathbf{x}(0) = 0$, what is the probability of reaching $\mathbf{x}^{(i)}$ using the asynchronous dynamics of equation (5)? What are the mean and variance of the convergence time?

(c) Starting again from the all-zero state, where does the synchronous dynamics lead?

Solution:

(a) $H[\mathbf{x}^{(i)}] = -z_i$. There are N states adjacent to $\mathbf{x}^{(i)}$. The all-zero state is one of these, with $H(\mathbf{0}) = 0$. The other N-1 are of the form $\mathbf{x}^{(i)} + \mathbf{x}^{(j)}$, $j \neq i$, and the energy is $C-x_i-z_j$. Since C is large in relation to the z's, all states adjacent to $\mathbf{x}^{(i)}$ have higher energy.

(b) Using the Glauber dynamics, the state evolves from $\mathbf{0}$ to $\mathbf{x}^{(i)}$ in one time step, where $i = n(0)$. Thus the absorption probability is 1/N for each i. The mean and variance are one and zero iterations, respectively.

(c) With synchronous dynamics, the network oscillates between the all OFF (zero) state and the all ON ($111...1 = 1^N$) state. Since $H(1^N) = N(N-1)C/2 - \Sigma_i z_i >> 0$, this motion is up-and-down in energy.

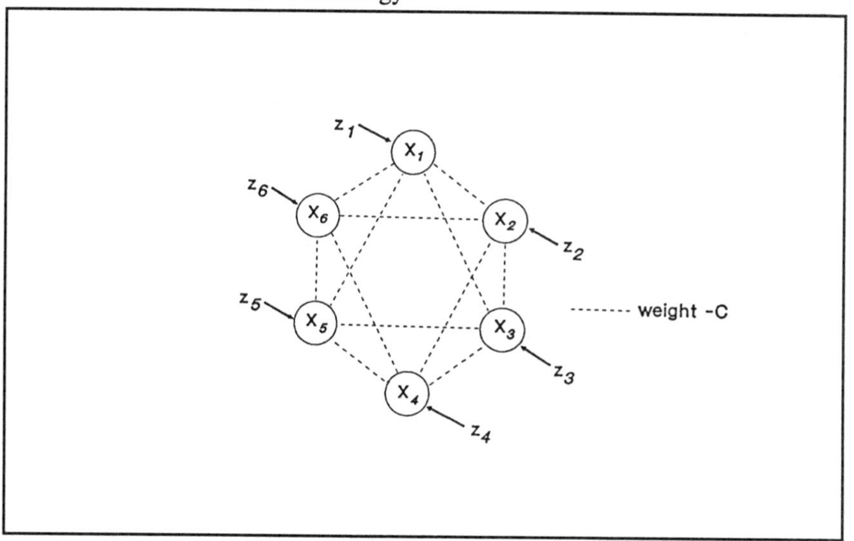

Network Diagram for N = 6

3.13 The weight matrix for this chain of N McCulloch-Pitts neurons is:

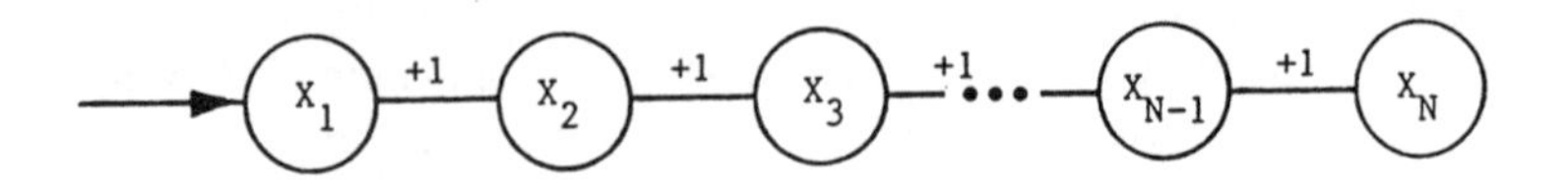

$$W_{ij} = \begin{cases} +1 \ \textit{if} \ |i-j| = 1 \\ 0 \ \textit{otherwise} \end{cases}$$

Let $\mathbf{x}(0) = \mathbf{0}$. Beginning at time $t = 0^+$, a positive stimulus is applied to neuron #1. Obviously the network state evolves from $00...0 = 0^N$ to $11...1 = 1^N$.

(a) How many synchronous sweeps does it take?

(b) What is the mean convergence time in sweeps of the asynchronous dynamics?

Solution: N synchronous sweeps[2]: $0^N \rightarrow 10^{N-1} \rightarrow 1^2 0^{N-2} \rightarrow ... \rightarrow 1^N$. (b) N random sweeps. Proof: The transition $1^n 0^{N-n} \rightarrow 1^{n+1} 0^{N-n-1}$ takes k+1 steps with probability $p_k = N^{-1}(1 - N^{-1})^k$, $k = 0,1,2,...$, for every n. This is a geometric distribution and the mean number of steps is:

$$1 + \sum_k k p_k = 1 + (N-1) = N \ .$$

The N stages of the evolution require a mean time of N^2 steps, or N sweeps of the network.

3.14 Let N McCulloch-Pitts neurons be linked by weights of +1: $W_{ij} = 1 - \delta_{ij}$. The only stable states are trivially the all ON and the all OFF. Let $\mathbf{x}(0) = 10^{N-1}$.

(a) How many asynchronous sweeps are required for convergence?

---

[2] The notation $1^2 0^4$ means 110000.

(b) What is the expected value of the number of sweeps to reach the all ON state? Give a numerical answer for N=6.

Solution:

(a) Two synchronous sweeps: $10^{N-1} \rightarrow 01^{N-1} \rightarrow 1^N$.

(b) A total of N-1 state transitions occur in mean time $m_1 + m_2 + \ldots + m_{n-1}$, where $m_k = (1\text{-}k/N) + 2(k/n)(1\text{-}k/N) + 3(k/N)^2(1\text{-}k/N) + \ldots = 1/(1\text{-}k/N)$. For N = 6, the mean convergence time is 13.7 steps = 2.3 sweeps. However, there is a 1/N chance of the transition to the all OFF state at the first step.

3.15 A student writes a program to find, by exhaustive search, all the stable states of a McCulloch-Pitts network with random, symmetric **W**, taking $W_{ii} = z_i = \theta_i = 0$ for every unit. The upper triangle consists of independent, zero mean random numbers. A computer virus, the infamous "Factor-of-Two Bug," modifies a program line that was to have symmetrized the matrix, causing the lower triangle to be **twice** the upper: $W_{ji} = 2W_{ij}$ for all $i<j$. What effect does the virus have on the computed results?

Solution: None. With reference to equation (3a), assuming asynchronous dynamics:

$$H = -(½)(\sum_{i,j} W_{ji}x_i\Delta x_j + \sum_{i,j} W_{ij}x_j\Delta x_i)$$
$$= -(3/2)u_i\Delta x_i$$

4. Analogs of Collective Computation

Most experts will agree that the study of neural networks is intrinsically multidisciplinary. Rosenblatt (1961), a psychologist, explained that

> It [was] only after much hesitation that [he] reconciled himself to the addition of the term "neurodynamics" to the list of such recent linguistic artifacts as "cybernetics", "bionics", "autonomics", "biomimetics", "synnoetics", "intelectronics", and "robotics."

Rosenblatt referred to "nerve nets" (his quotes) but never to "neural networks."

Leo Szilard, who (with Fermi) split the atom, and talked Einstein into writing to Roosevelt in behalf of The Bomb, took up the study of neurobiology in 1948, and, in his last year (1964), published a paper that described "neural networks" (his quotes) using a diagram like those in the present text. It would seem that the term "neural networks", so common since its use by Hopfield (1982), owes much of its popularity to the writings of physicists.

Multidisciplinarity is no less characteristic of collective computation in symmetric networks than it is of neural networks in general. This chapter briefly examines the phenomenon from the perspectives of the physicist and the electronics engineer.

Spin Glass Models

Spin glass models form the bridge between traditional physics and the study of collective computation in neural networks. Their ancestry is traced to Ernst Ising's 1925 doctoral dissertation. The Ising model concerns the physics of phase transitions, the large-scale, qualitative changes in the state of matter interposed between small changes in parameters like pressure and temperature. The purpose of Ising's thesis was to understand ferromagnetism, especially the phenomenon of "spontaneous magnetization."

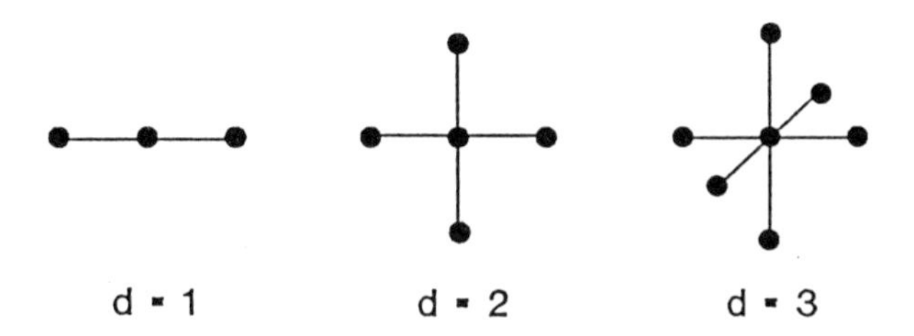

Following Cipra (1988), the formulation of the model begins with the notion of a lattice, understood as a finite set of regularly spaced points in a space of dimension d = 1, 2, or 3.

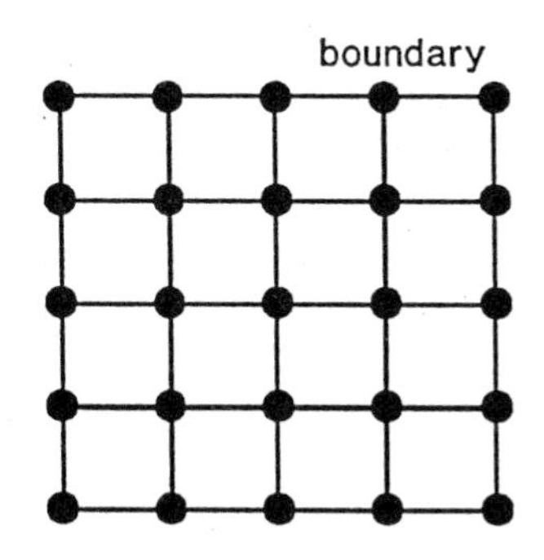

Lattice for d = 2.

Each segment drawn between lattice sites is called a bond; and the sites are nearest neighbors if and only if there is a bond between them. Except for sites on the boundary of the lattice, a site has 2d nearest neighbors.

In the Ising model of ferromagnetism, lattice sites are occupied by magnetic atoms whose dipole moments are pointed either up (+1) or down (-1). Therefore an independent variable, s_i, may be assigned to each site: i = 1, 2 ,..., N. These variables can assume two values, $s_i = \pm 1$, called the states of the site. Since the dipole orientation is determined by the *spin* of an atomic particle, s_i can also be regarded as giving the directional sense of the ith spin. An assignment $s = (s_1,\ldots,s_N)$ of sign/spin/orientation is called a *configuration* of the system.

Now form the *Hamiltonian* (or total energy) of the system. Since the field of a dipole follows the inverse-cube law, assume that only the short-range, nearest neighbor interactions, and interactions of the lattice sites with an external field, contribute to the total energy. For each configuration there is an energy:

$$H(s) = -\sum_{<i,j>} J s_i s_j - \sum_i K s_i \ ,$$

where J is the interaction coefficient that gives the strength of the nearest neighbor bond, K is the external field, and the summation over $<i,j>$ includes only nearest neighbor pairs. Since magnetic dipoles attract each other when they are oriented in parallel with each other, the interaction coefficient is $J > 0$.

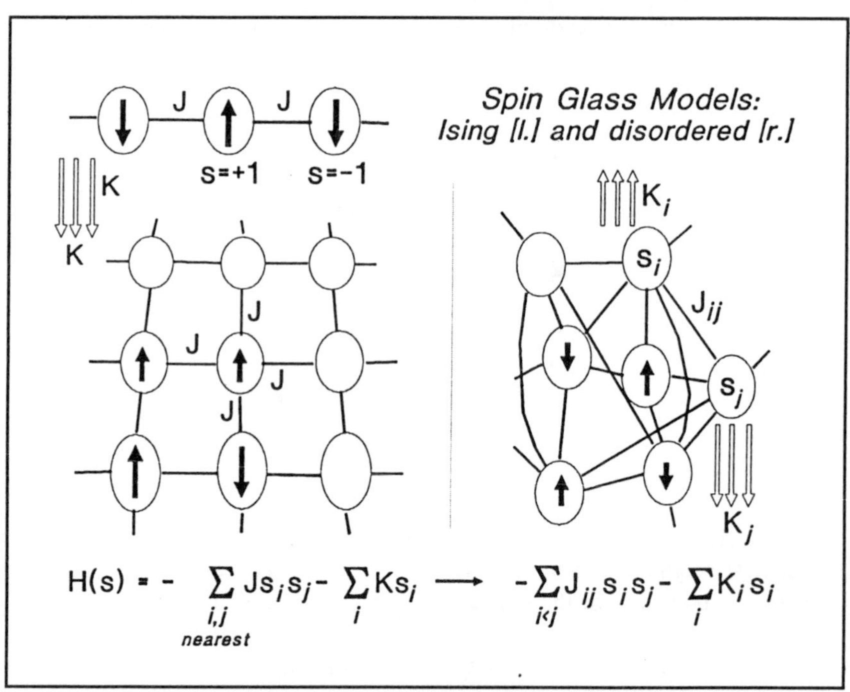

Figure 4.1 Spin Glass Models: Ising (left) and disordered (right).

Recalling the definition of *nearest neighbors*, proximity is defined by connectivity. All N sites can be neighbors if they are bonded together. In a physical system, constrained to three dimensions, such bonding can be imagined as the result of placing ribbons of a magnetic conductor (say, soft iron) between pairs of dipoles. The thickness of the soft iron ribbon from site i to site j determines $|J_{ij}|$. The presence or absence of a half-twist decides sign(J_{ij}). Clearly $J_{ij} = J_{ji}$ as in free space; and $J_{ii} = 0$. This flight of fancy leads to the spin glass Hamiltonian:

$$H(s) = -\sum_{i<j} J_{ij}\, s_i\, s_j - \sum_i K_i\, s_i \qquad (1)$$

after routing a ribbon from an external source to make K_i site-specific. Fig. 4.1 illustrates this generalization of the Ising model. Any doubt that H(**s**) is indeed an energy function should be dispelled by the substitution of (**x**,**W**,**z**) for (**s**,**J**,**K**), which returns us to the equation (3.2). Problem 3.1 showed that sign neurons (Ising spins), like McCulloch-Pitts neurons, collectively act to minimize the global energy. In the language of physics, the *net input* to a neuron:

$$Y_i = \sum_j W_{ij}\, x_j \quad ,$$

corresponds to the *local field* (usually $h_i = \sum_j J_{ij} s_j$).

Problem 4.1: Fig. 4.2 shows three configurations of an Ising model with N = 64 and d = 2. The nearest neighbor bonds of strength J = +1 are not drawn explicitly. Assume K = 0. Compare their energies. Which configuration is stable?

Solution: The energies are (fig. 4.2a) H = -112 and (fig. 4.2b) H = -96. Both are stable, since the first case is the absolute minimum energy configuration, and the flip of a spin on the borderline between the two "domains" of the last case would *increase* the energy by ΔH = 2 (fig. 4.2c).

Returning to the Ising model of ferromagnetism, the phase transition occurs with the appearance of spontaneous magnetization. At any constant, nonzero temperature, the external field will induce magnetization in the lattice, as the dipoles align themselves with it, causing the energy to decrease. If the field is slowly reduced to zero, the behavior of the lattice sites depends critically on temperature. Above a critical temperature, the lattice becomes disordered and the bulk magnetization, which is proportional to $\sum x_i s_i$, vanishes. Below the critical temperature, there is a residual (or spontaneous) magnetization as a majority of lattice sites "remember" the behavior imposed on them by the field.

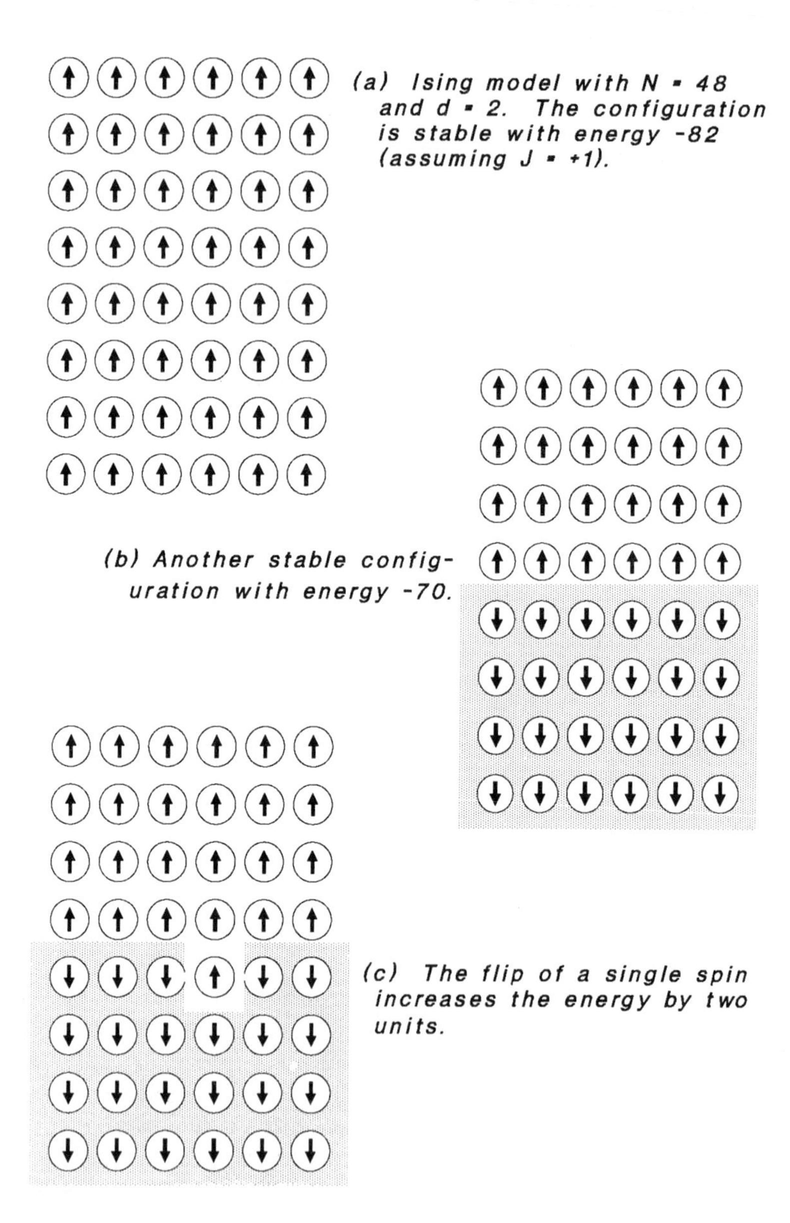

(a) Ising model with N = 48 and d = 2. The configuration is stable with energy -82 (assuming J = +1).
(b) Another stable configuration with energy -70.
(c) The flip of a single spin increases the energy by two units.

Figure 4.2 Configuration of an Ising model.

The fact that the one-dimensional model does *not* exhibit a phase ansition at any temperature discouraged Ising from pursuing the subject. ıbsequent research has shown the existence of the transition for d = 2. See ramers & Wannier (1941), page 252. Recall Hopfield (1982) that a suitable ntent-addressable memory could retrieve this information from a partial intout of the citation. The three dimensional model has yielded no analytic lutions (Cipra, 1988).

By exponentiating the scaled Hamiltonian and summing over all 2^N spin nfigurations, one obtains the *partition function:*

$$Z = Z(\beta, J, K, N) = \sum_{s} \exp[-\beta H(s)] \quad . \tag{2}$$

ıe scale factor is $\beta = 1/kT$ in the physical system for which k is Boltzmann's nstant and T is the absolute temperature. The temperature is a manifestation random kinetic activity — the pushing and pulling of lattice sites by one ıother and by the "heat bath" in which the whole system resides. This random tivity works against the establishment of the short-range order that (together ith long-range disorder) characterizes the glass.

Problem 4.22: Find closed-form expressions for the partition functions the two-dimensional Ising models with K = 0 and N = 2 and 3.

Solution: $Z = 4\cosh(\beta J)$ and $8\cosh^2(\beta J)$ for N = 2 and 3, spectively.

Problem 4.3: Change the statement of the preceding problem to ipulate that the lattice sites are occupied by McCulloch-Pitts neurons and write e partition function for N=2.

Solution: $Z = 3 + \exp(\beta J)$.

Ackley, Hinton, and Sejnowski (1985) proposed simulated annealing to slodge the Hopfield net from local energy minima and enable it to settle into ates of still lower energy that would represent better (if still suboptimal) lutions to the problem thus represented. The network is "heated" by the ldition of noise to the bias (or threshold) of each unit. When these noises are dependent, identically distributed random variables, the network state (**S** or **X**) kes a walk on the vertices of a unit N-cube. The long-run (limiting or ationary) distribution:

$$Pr(S = s) = Z^{-1}\exp[-\beta H(s)] \quad , \tag{3}$$

formally equivalent to the equilibrium distribution of states in a spin glass, ith Z being the partition function (equation 2). The assertion of Ackley, inton, and Sejnowski, that $T = (1/\beta)$ is proportional to the root mean intensity of noise described by a logistic distribution, was not powerfully motivated. Shaw et al. (1979) had earlier arrived at a similar expression in which β is a "smearing factor" determined from details of a stochastic model of the chemical synapse.

It was over a hundred years ago that Gibbs sought time-invariant solutions to a Liouville equation in which the independent variables were the (Hamiltonian or phase space) coordinates of a multiparticle system and the dependent variable was the probability of the system being in a given state (or configuration). He arrived at a canonical ensemble in which "the index of probability [i.e., the log-probability] is a linear function of the energy" of the state. This result is expressed by equation (3), equivalent to:

$$\beta^{-1}\ln(P) = -\beta^{-1}\ln(Z) - H \quad ,$$

and often called the Boltzmann distribution. Belief in the possibility of a mathematical treatment of biological intelligence, patterned after statistical thermodynamics, goes back at least as far as the works of John von Neumann (1954) — published posthumously. It is not surprising that this belief finds expression in recent studies of connectionist models. The validity of the Boltzmann distribution in the context of the neural network paradigm is solely dependent on the existence of models that give rise to it. A more detailed discussion of the stochastic model at nonzero temperature is the subject of chapter 6; and it will be motivated by the study of Hopfield-Tank optimization networks in the next chapter. Chapter 6 will also recall Glauber's (1963) "Time-Dependent Statistics of the Ising Model" in which physicists like Crisanti and Sompolinsky (1988) and van Hemmen et al. (1988) have identified the same dynamical assumptions that Hopfield (1982) based his stochastic model on.

Electronic Neurocomputers for Collective Computation

From the viewpoint of the electronics engineer, the collective computation network is an analog circuit in which nonlinear amplifiers feed back to each other through a symmetric array of coupling conductances. Fig. 4.3 shows this "crossbar circuit," in which the ith amplifier produces output voltage $V_i = F(\phi_i)$, where $\phi_i(t)$ is the time-dependent driving point voltage. The driving point of each amplifier is shunted to ground by a resistance R and a capacitance C in parallel as shown earlier (fig. 2.1). This RC circuit provides the inertia that would characterize any real amplifier (or neuron); and its use with ideal, delayless amplifiers makes the mathematical model conform better to the electronic circuit.

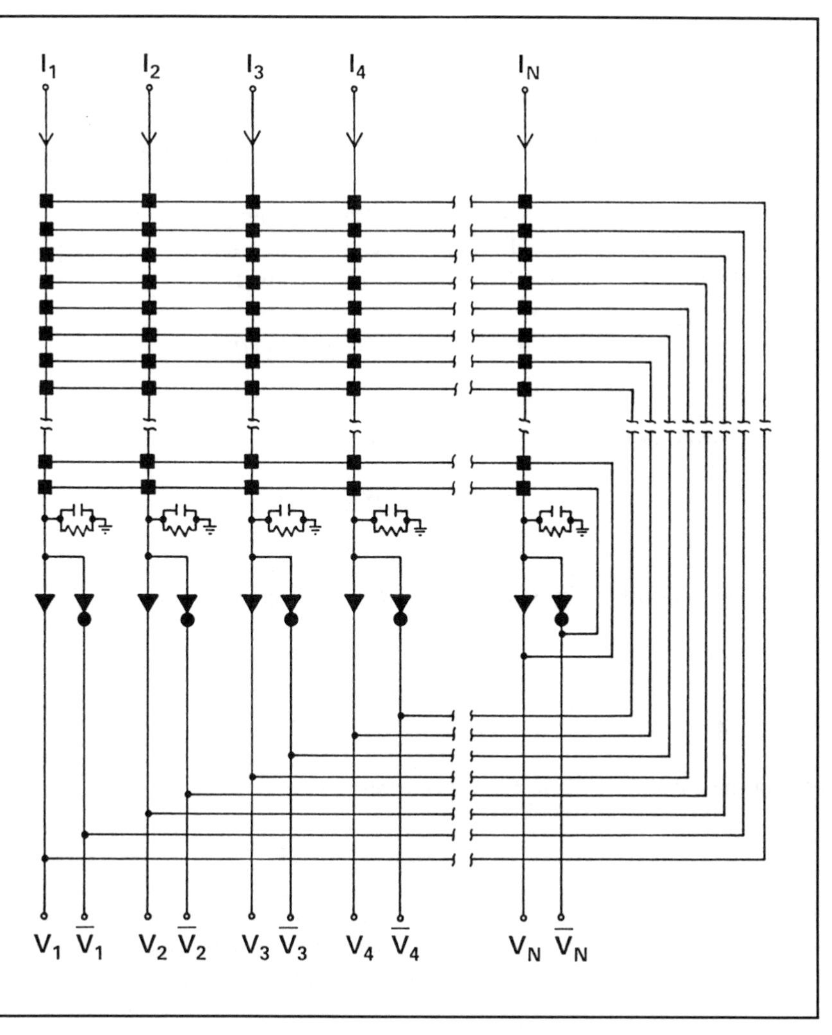

Figure 4.3 Fully analog collective computation network.

Each amplifier is paired with an invertor that outputs $\overline{V}_i = -V_i$. These outputs feed back to the driving points through resistances R_{ij}. The current injected by the jth amplifier at the driving point of the ith is:

$$y_{ij} = (V_j - \phi_i)/R_{ij} \qquad \textbf{(4a)}$$

if the noninverting output is taken, or:

$$y_{ij} = (V_j + \phi_i)/(-R_{ij}) \tag{4b}$$

otherwise. We write $T_{ij} = \pm 1/R_{ij}$ for the coupling conductance, the sign being determined by the connection. The assumption is that either (4a) *or* (4b) applies: No amplifier is connected to the same driving point as its invertor and vice versa. Then the sum of all the currents injected into the driving point of the ith amplifier by all the other amplifier/invertor pairs is:

$$y_i = \sum_j Y_{ij} = \sum_j T_{ij} V_j - \phi_i \sum_j (1/R_{ij}) \quad . \tag{5}$$

The argument accompanying fig. 2.1 shows:

$$Cd\phi_i/dt + \phi/R = y_i + I_i.$$

which, on multiplying by R to get the IR drop, becomes:

$$\frac{d\phi}{dt} + \frac{\phi}{R} = I_{Bias}$$

adding the current of (5) to the bias current gives:

$$Cd\phi_i/dt + \phi_i/R = y_i + I_i \ ,$$

the right side being the sum of (5) and the bias current I_{bias} (which is constant). This sum is the total current shunted to ground at the driving point. Substituting equation (5) for the net cross-coupled current and rearranging gives:

$$C\frac{d\phi_i}{dt} = -\phi_i \left[\frac{1}{R} + \sum_j \frac{1}{R_{ij}} \right] + \sum_j T_{ij} V_j + I_{Bias}. \tag{6}$$

To make the summing resistance (R) very small compared to the smallest coupling resistance, take:

$$\sum_j |T_{i\,j}|\; R <<1 \qquad \forall i \quad . \tag{7}$$

This would be achieved in practice by using transconductance amplifiers with very low input resistance. But if the available components are just garden-variety op amps with moderately high input resistances, we can follow Hopfield (1984) and define:

$$\frac{1}{R_i} = \frac{1}{R} + \sum_j \left(\frac{1}{R_{ij}} \right) \quad .$$

Then equation (6) reduces to:

$$C\frac{d\phi_i}{dt} = -\frac{\phi_i}{R_i} + \sum_j T_{ij}\; V_j + I_{Bias} \quad .$$

Multiplying through by R_i, this is:

$$\tau_i \frac{d\phi_i}{dt} = -\phi_i + \sum_j W_{ij}\; V_j + z_i \quad ,$$

in which $\tau_i = R_i C$ is a time constant, $W_{i\,j} = R_i T_{i\,j}$ is a dimensionless weight, and $z_i = R_i I_i$ is a bias voltage. The symmetry of **W** (again) makes this a special case of the Cohen-Grossberg system. In the high-gain limit, where the amplifiers become hard limiters, as $F(u) = \mathcal{H}(u)$ or $\mathrm{sgn}(u)$, the stable points are local minima of:

$$-\sum_{i<j} W_{ij}\; V_i\; V_j - \sum_i V_i\; z_i \quad .$$

Moreover, if condition (7) pertains, this expression can be divided by R to read:

$$E = -\sum_{i<j} T_{ij}\; V_i\; V_j - \sum_i V_i\; I_i \quad .$$

This computational energy has the physical dimensions of (current times voltage equals) power. It is not the same, however, as the power dissipated by the circuit. The power would be obtained, in the absence of biases, as the sum over all the amplifiers (and invertors) of the output voltage times the output current, roughly $\Sigma_i \Sigma_j (V_i)^2 |T_{ij}|$. The computational energy, in the absence of biases, is -½ the sum of all the output voltages times the input (driving point) currents:

$$E = -\tfrac{1}{2} \sum_i V_i Y_i \approx -\tfrac{1}{2} \sum_i V_i \left(\sum_j T_{ij} V_j \right) = -\sum_{i<j} T_{ij} V_i V_j.$$

Fig. 4.4 is an alternative depiction of the analog neural network emphasizing its bus architecture instead of the crossbar structure. (Both circuit diagrams, figs. 4.3 and 4.4, are sufficiently general to encompass all kinds of connectionist models, weight matrix symmetry being a special case.) Instead of a square array of resistors, this sketch features a modular structure in which each neuronlike amplifier is fed by a "synapse chip" making N (or N-1) connections to the N-line bus. Invertors are placed at the synapses. Each "chip" contains a *row* $\mathbf{T}^n$ of the conductance matrix.

The modularity of this network is subverted by the fact that the bus lines proliferate with the neurons. The state of the art in electronic neurocomputer design is probably reflected in the proceedings of a recent International Joint Conference on Neural Networks, in which only one paper (Mueller et al., 1989) embraces the fully analog design approach. That design uses analog multiplexers to route the outputs of the neuronlike amplifiers back to the synaptic inputs. While this is motivated by sheer necessity, and the understanding that hundreds of units will have to be employed to solve the smallest practical problems, the dynamics of the multiplexed circuit would seem to have escaped a rigorous mathematical treatment thus far.

Putting aside the uncertainties pertaining to its fabrication, an important point in favor of the analog collective computation network is its *speed*. Fig. 4.5 compares the digital simulation model (left) with the analog circuit (right). Assuming discrete, asynchronous dynamics, the former must perform one multiply-and-accumulate operation, or inner product step (i.p.s.), for each of the $N-1 \approx N$ synapses every time a neuron is selected for update. Thus a *sweep* of the net involves about N^2 i.p.s. (or *interconnects);* and the convergence time grows with network size as the number of sweeps times $N^2\Delta t$, where $1/\Delta t$ is the flop (floating point operation) rate.

The essential point of Hopfield (1984) was that the stable states of the analog circuit, in the high-gain limit, are the same as those of the stochastic model. The relationship between convergence time in the two models was not explored. Hopfield and Tank (1986c) suggested that convergence time relative to the RC time constant of the circuit is proportional to sweeps of the stochastic model. The heuristic argument must be as follows:

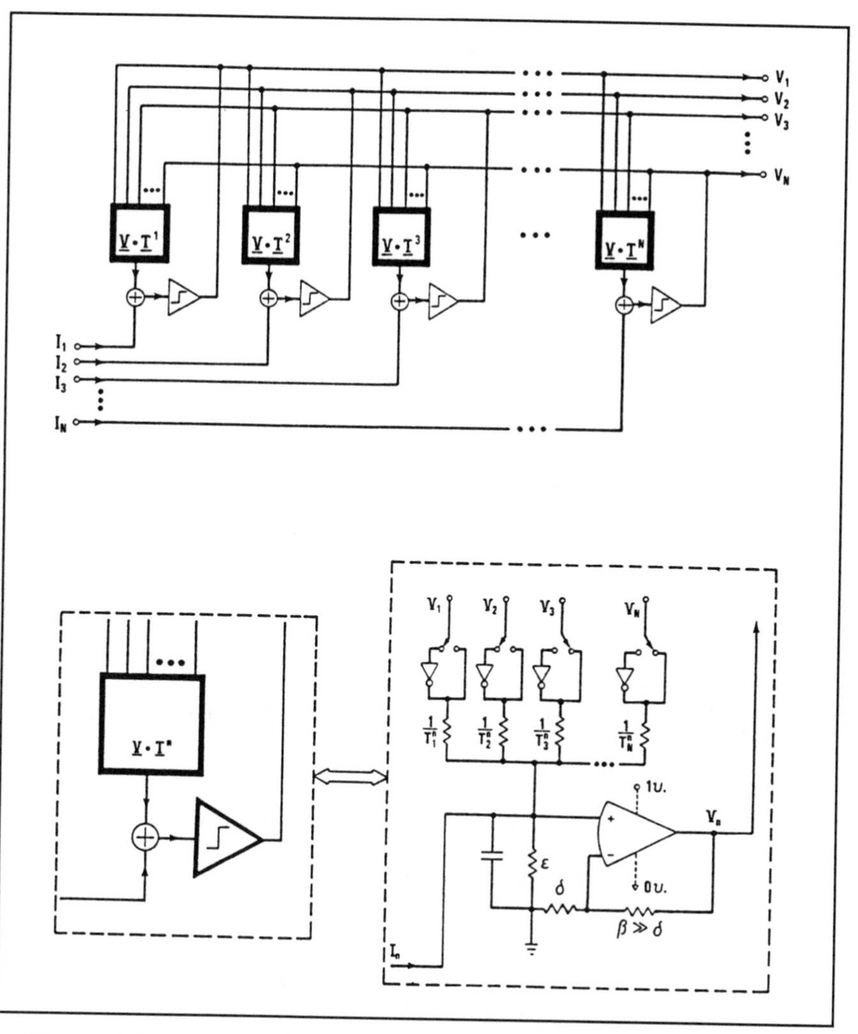

Figure 4.4 Neural network depiction emphasizing bus architecture.

The time required for a high-gain amplifier to toggle is proportional to RC. Therefore the number of state changes experienced by an electronic neuron in the course of a given evolution grows (at least initially) as t/RC. The number of state changes experienced by a binary unit in the stochastic model is on the order of one per sweep. Thus the ratio of the real convergence times is approximately $N^2\Delta t/RC$, that for the digital simulation model being in the numerator.

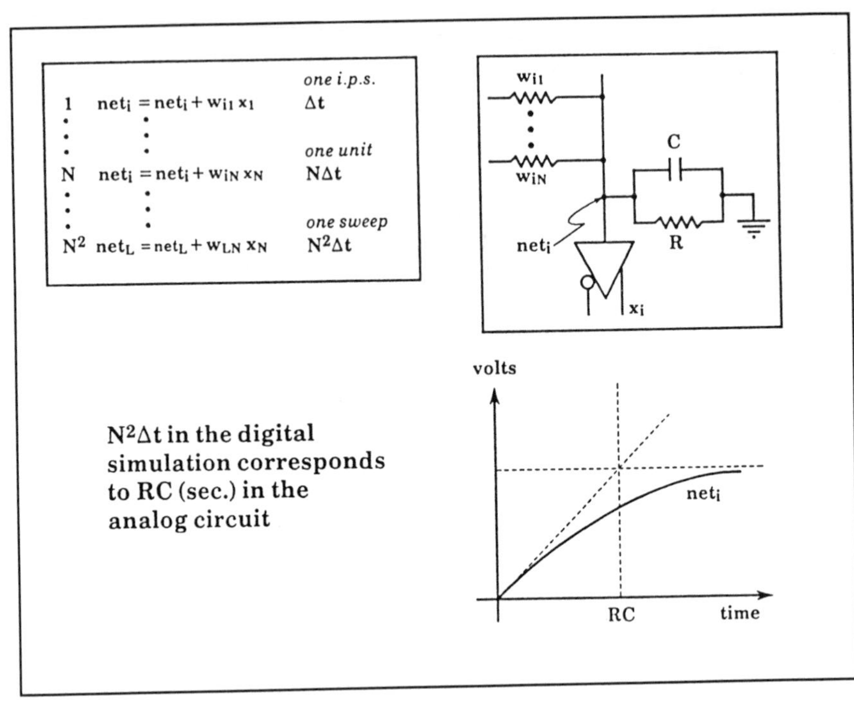

Figure 4.5 Comparing convergence time - digital vs. analog.

Table I Median Convergence Time: Digital vs. Analog

median sweeps = $\mu(N)$ = $\sqrt{(N/8)}\ell n(2N)$

N	$\mu(N)$	Δt	$\mu(N)N^2\Delta t$	$\mu(N)RC$
100	8	10^{-6}	80 ms	80 ms
10^3	31	10^{-7}	3.1 s	0.31 s
10^4	157	10^{-8}	157 s	1.57 s
10^{10}	1.7×10^5	10^{-9}	1.7 B yrs.	1700 s

N = network size 1/RC = 100 Hz

Table 1 compares median convergence times for the digital model (fourth column) and the analog circuit (fifth column) for selected N and Δt, assuming a time constant of RC = 10 millisec. With N = 100, a serial simulation that executes a million inter-connects per second converges as quickly as the analog circuit. At the other extreme, with the units about as numerous as our cerebral neurons, there is a tremendous disparity. These calculations assume that the convergence time, in sweeps or in time constants, tends to grow slightly faster than $N^{1/2}$, as was claimed in chapter 3, page 64. On the other hand, if the median number of sweeps to convergence of the digital model is the same irrespective of network size, and if the heuristic argument just offered is valid, then the fifth column of the table would have to show 80 msec to convergence — until N grows so large that propagation delays, long line inductances, and other unwanted effects conspire to invalidate the modeling assumptions.

5. Shaping the Energy Landscape for "Good, Suboptimal Solutions"

Having shown how collective computation networks (CCNs) with stochastic dynamics act as content-addressable memories when patterns are stored using the outer product rule, Hopfield (1984) reexamined the network assuming the continuous, deterministic dynamics of the analog circuit model. Subsequently Hopfield and Tank (1986a,1986c) exhibited "neural circuits" for solving optimization problems. As before, they cited the logical calculus of McCulloch and Pitts (1943) and the perceptrons of Rosenblatt as prior efforts. This time, however, they rejected both the global clock (which never really existed in perceptrons) *and* the assumption of binary units. We break with them here in so far as our discussion will continue to concern binary units in symmetric nets using stochastic (Glauber) dynamics. This break will open some difficult questions of comparability when we analyze the performance of CCNs for constraint satisfaction and optimization. On the other hand, our dynamical assumptions will guarantee that the energy landscapes have essentially the same local minima and (hence) stable states, which are one-to-one with the hypothetical solutions that the net may offer for the given problem.

Rosenblatt (1961) drew a distinction between the *monotypic* and *genotypic* approaches to neural networks. The monotypic was epitomized by McCulloch and Pitts, who showed (as it were) that a special-purpose neural circuit could, in principle, be crafted to perform any kind of logical calculation whatever. From 1957 on, the research trend followed Rosenblatt's genotypic approach, seeking to find what capabilities and limitations pertained to nets of a given architectural class subject to certain general rules for determining the weights. The optimization networks of Hopfield and Tank synthesize both approaches, the genre being delineated by weight matrix symmetry and the detailed "logic" being embodied in the particular weight matrix with reference to the energy landscape it engenders.

Data Representation

The first step in designing a neural optimization circuit is to map the mathematical problem into a collection of neuron like units by establishing a itable, distributed representation of the variables that define possible solutions, ntrol inputs, and constraints.

Example 1 (Analog-to Digital [A-to-D] converter): fig. 5.1 illustrates is process for the 4-bit A-to-D converter (Hopfield and Tank, 1986a,1986c).

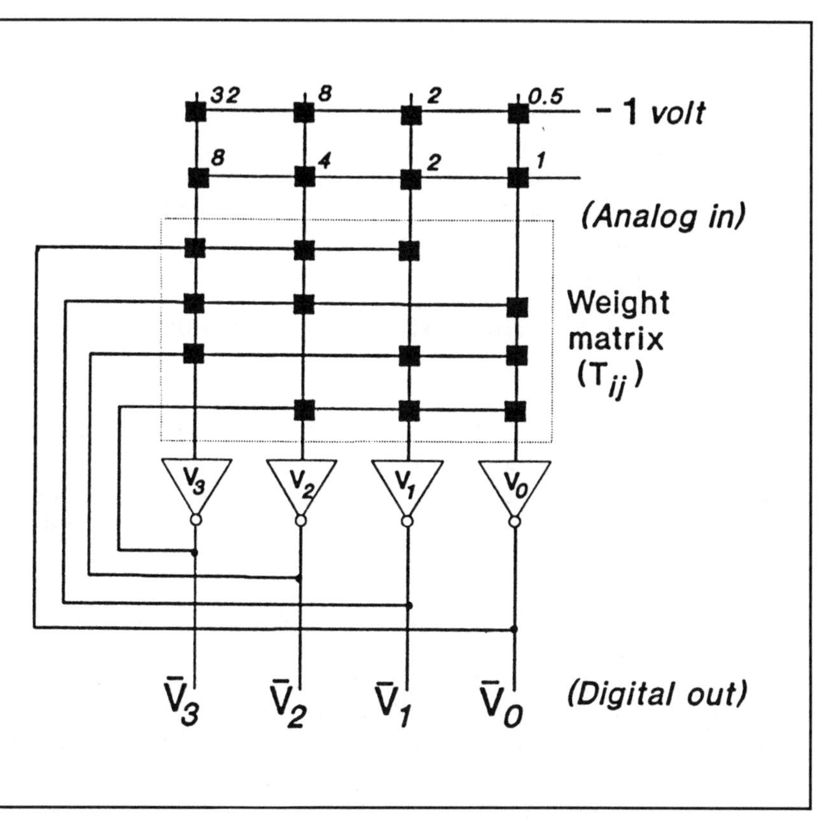

Figure 5.1 Four bit analog-to-digital converter.

he figure uses the TVI notation (chapter 1, table 1) in which $V_i \in \{0,1\}$ is the ate of the ith McCulloch-Pitts neuron. The base two number $V_3V_2V_1V_0$, rresponding to network state $\mathbf{V} = (V_0,\ldots,V_3)$, is the digital output of the net. ne input being a real voltage ψ, the problem is to determine the weights (T_{ij}) d biases (I_i) that will make the network converge to stable states such that:

$$\sum_{i=0}^{3} V_i \, 2^i = \psi$$

to the extent allowed by the digitization. The figure already exhibits part of t
solution, since the eight resistors (represented by filled squares), who
conductances are given, establish the bias currents. For example, $I_3 = 8 \psi$
32.

Example 2: The traveling salesman problem (TSP) is to find a path
minimum distance connecting M cities. The path (or tour) has M legs, since t
city of origin must also be the final destination. The cities are numbered by
{1,...,M} and D(i,j) is the distance between city i and city j. The first stop
at city j[1] and so on until the last stop at city j[M] = j[0], the city of origi
The objective is to minimize:

$$D_T = \sum_{i=0}^{M-2} D\,(j[i], j[i+1]) + D\,(j[M-1], j[0]) \;.$$

With reference to fig. 5.2, TSPs come in two kinds, Euclidean and no
Euclidean. In the first case, there exists a map from which all interci
distances can be determined with the aid of a ruler. In the second case, it
necessary to represent **D** as a symmetric MxM matrix with all zero diagon
elements.

Fig. 5.2 compares the map representation of an optimal five-ci
Euclidean tour with a digital representation consisting of indicator lamps driv
by a neural circuit. The ON/OFF indicators form a 5x5 array and are in one-t
one correspondence with the McCulloch-Pitts neurons. The representati
shown in the figure suggests that the network state will be a square matrix:

$$V_{ij} = \begin{cases} 1 & \textit{if the stop } i-1 \textit{ is at city } j \\ 0 & \textit{otherwise} \end{cases} ,$$

i,j = 1, 2, ..., M, instead of a vector string of $N = M^2$ components. The sta
space of the network is identified with the 2^N corners of an M^2-cube; but t
solution space of the problem thus represented includes only (M-1)! of the
points.

The data representation determines the number of units and establish
the boundaries that partition state space into regions distinguished by whether t
points in them are valid solutions or not. In the case of the A-to-D convert
the $2^4 = 16$ possible states all belonging to the solution space. For the TSP

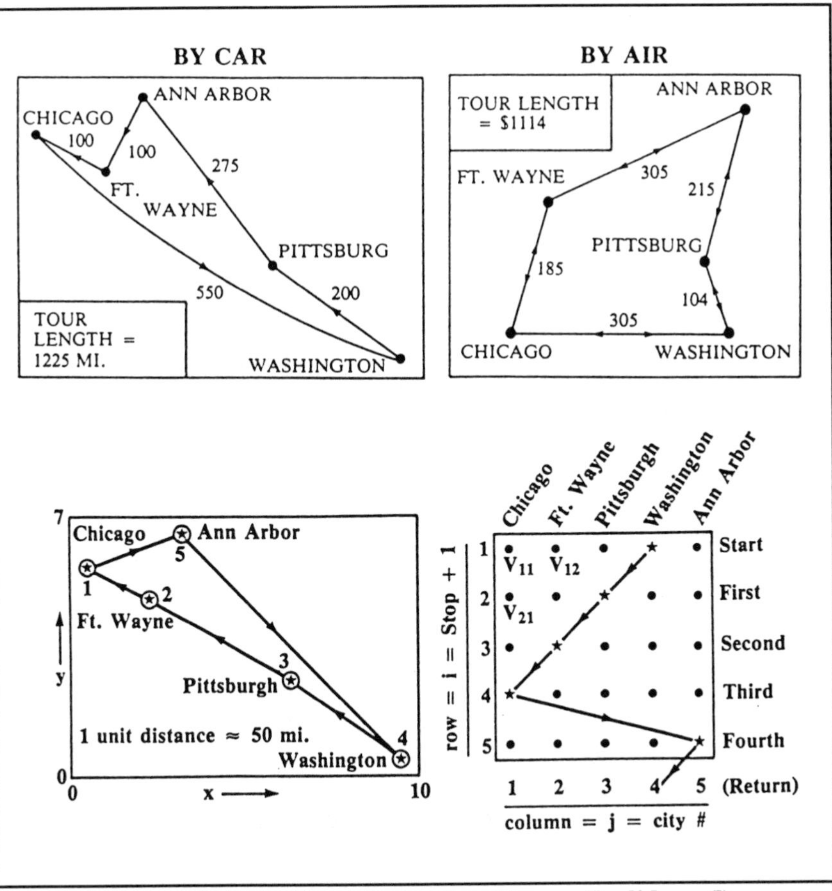

Figure 5.2 Traveling salesman problems: Euclidean (l) and non-Euclidean (r).

circuit, the number of solutions, (M-1)!, is, even for moderate M, a tiny fraction of the total number of states 2^{M^2}; and:

$$(M-1)!\exp(-M^2) \rightarrow 0$$

as the number of cities grows.

Shaping the Energy Landscape

The remaining agenda is to place the stable states into correspondence with the valid solutions, by creating energy minima at their coordinates, and to further shape the energy landscape so that better solutions are preferred by the network. Given a random $\mathbf{V}(0)$, we want $\mathbf{V}(t)$ to be absorbed into a stable state $\mathbf{V}^o$ in the solution space with greater or lesser probability depending on QUALITY($\mathbf{V}^o$). In general, we can formulate the problem so that:

$$\text{QUALITY} = \text{VALUE} - \text{COST} - \text{PENALTY}, \qquad \textbf{(1a)}$$

where value is what the solution gains us, cost is taken literally, and the penalty is assessed on solutions that lack the desired clarity or integrity. Ideally, we would like an energy landscape that is so dominated by the valleys surrounding the cleverest solutions that we need not trouble with the question mark in:

$$Q_{\omega ?} = \Pr[\text{absorption by } \mathbf{V}^o = \omega \mid \text{initial state } \mathbf{V}(0) = ?].$$

We design the network so that:

$$\text{ENERGY} + \text{CONSTANT} = -\text{QUALITY}. \qquad \textbf{(1b)}$$

The basins of attraction of the stable states will then be directly proportional *in depth* to solution quality. This technique will be effective to the extent that greater depth and greater longitudinal extent (i.e., the size of the basin of attraction) are concomitant. Combining equations (1a) and (1b) gives the design identity:

$$\text{ENERGY} + \text{CONSTANT} = -\text{VALUE} + \text{COST} + \text{PENALTY} . \qquad \textbf{(1c)}$$

Problems Involving Quadratic Cost

A number of optimization networks can be regarded as special cases of the following identifications:

$$VALUE(V) = \sum_{i=1}^{N} a_i V_i \quad , \qquad \textbf{(2a)}$$

$$COST(V) = \sum_{k=1}^{k} d_k (b_k - \sum_{i=1}^{N} C_{ki} V_i)^2 \quad , \tag{2b}$$

nd

$$PENALTY(V) = \sum_{i=1}^{N} e_i V_i (1 - V_i) \quad . \tag{2c}$$

'he state of the ith neuron is V_i, $0 \leq V_i \leq 1$. The coefficient a_i measures the alue of having the ith unit fully ON and the coefficient C_{ki} is the corresponding ost relative to the kth criterion. The COST function is a weighted sum over all criteria of the squared deviations of the costs relative to target values $\{b_k\}$. he PENALTY, for $e_i > 0$, is assessed against all units that equivocate by iling to settle to one of the endpoints of the unit interval. This PENALTY rm is germane to circuits with graded neuron/amplifiers that need to be forced to saturation. When McCulloch-Pitts neurons are assumed, the PENALTY rm (2c) vanishes. Once again, the ENERGY function is:

$$E(V) = -(\tfrac{1}{2}) \sum_{j=1}^{N} \sum_{i=1}^{N} T_{ij} V_i V_j - \sum_{i=1}^{N} I_i V_i \quad . \tag{2d}$$

ubstituting equations (2a) through (2d) into (1c) and rearranging, we find that:

$$CONSTANT = \sum_{k=1}^{K} d_k (b_k)^2 \quad , \tag{3a}$$

$$I_j = a_j + \sum_{k=1}^{K} d_k c_{kj} (2b_k - C_{kj}) \quad , \tag{3b}$$

nd

$$T_{ij} = -2 \sum_{k=1}^{K} d_k C_{kj} C_{ki} \quad , \tag{3c}$$

o long as:

$$e_j = \sum_{k=1}^{K} d_k (C_{kj})^2. \quad \textbf{(3d)}$$

Equation (3c) assumes i $\neq$ j; and (3d) guarantees that $T_{ii} = 0$.

Example 1, ctd.: The cost function of the four-bit A-to-D converte circuit is the squared error

$$COST(V) = \tfrac{1}{2} \left[\psi - \sum_{j=0}^{3} V_j 2^j \right]^2$$

in which the neuron index runs from zero to N-1. Noting this discrepancy an rewriting the penalty function (2c) accordingly, we set VALUE = 0. Then th problem is a special case of equations (2) in which $\mathbf{a} = 0$, K = 1, b = ψ $C_{kj} = 2^j$, and $d_k = 1/2$. With reference to equations (3), the design solution $I_j = 2^j(\psi - 2^{j-1})$ and $T_{ij} = -2^{i+j}$. Substituting these results in equation (2d) an setting $\psi = 2$ gives:

$$E(V) = \sum_i \sum_j 2^{i+j-1} - \sum_i 2^i (2 - 2^{i-1})$$

for the input-specific energy landscape. The state $V_3V_2V_1V_0 = 0010$ (which a binary representation of 2) has energy -2. This is indeed a stable state, sinc each of the four adjacent states (obtained by toggling a single unit) lies uphill i energy.

Problem 5.1: The 0-1 knapsack problem is to maximize $\sum_{i=1}^{N} a_i$ subject to $\sum_{i=1}^{N} C_i V_i \leq b$. The variable V_i is the 0-1 indicator of whether the out of N items is selected. Assume a_i, C_i, and $b > 0$. Design a collectiv computation network to solve this problem.

An equivalent statement of the problem is to maximize:

$$QUALITY(V) = (\sum_{i=1}^{N} a_i V_i) \ \mathcal{H} (b - \sum_{i=1}^{N} C_i V_i)$$

in which $\mathcal{H}(.)$ is the familiar step function. Because this is *not* a quadrat form, the solution in terms of a CCN seems elusive.

The next problem can be viewed as a modification of the 0-1 knapsack problem in which a quadratic cost function replaces the step function.

Problem 5.2: Richard Greenback, MBA, has just received $20 million from his uncle for use in making strategic investments in Eastern Blocks real estate. He acquires a list of properties $i=1,\ldots,N$ each having a current price C_i and an estimated resale value a_i ten years hence. Since Rich's credit is good, and his uncle trusts him implicitly, it is only necessary for the total investment to be $20 million $\pm$ 10%. Devise a collective computation network to solve this problem.

Solution: The data representation is a straightforward assignment of one McCulloch-Pitts neuron to each property. The V_i is the 0-1 indicator of whether the ith property is purchased. The solution space consists of all $\mathbf{V}$ for which $\Sigma_i C_i V_i \leq b$. Let:

$$COST(\mathbf{V}) = (b - \sum_{i=1}^{N} C_i V_i)^2 d$$

and:

$$VALUE(\mathbf{V}) = \sum_{i=1}^{N} a_i V_i$$

with PENALTY = 0. If the total investment differs by 10% from the budget b = $20 million, then COST = $b^2 d/100$. Making d very large will raise tall barriers against solutions that are off target. Some relation such as $d \approx (100\ \Sigma_i a_i)/Nb^2$ will probably constrain the network to stay within the allowed error bounds most of the time. With reference to equation (3), the network is defined by:

$$I_i = a_i /d + 2bC_i - (C_i)^2$$

and:

$$T_{ij} = -2C_i C_j$$

after dividing both $\mathbf{I}$ and $\mathbf{T}$ by $d > 0$, which affects neither the fixed points nor their basins of attraction.

Problem 5.3 (Signal Decomposition): A real-valued, band-limited signal $S(t)$ is sampled at times $t = 1,\ldots,K$. Devise a neural circuit to decompose the signal into a sum of basis functions of the form $g(t;\mu,\sigma)$, where the parameters are $\mu = 1, \ldots, M$ and $\sigma = 1,\ldots,L$.

Solution: Following Hopfield and Tank (1986a), minimize the sum of the squared errors:

$$COST(V) = \sum_{t=1}^{K} [S(t) - \sum_{\mu} \sum_{\sigma} V_{\mu\sigma} g(t;\mu,\sigma)]^2 \quad \textbf{(4)}$$

where the coefficients of the expansion, $V_{\mu\sigma}$, lie on the interval [0,1].

Let:

$$\hat{S}(t) = \sum_{\mu} \sum_{\sigma} V_{\mu\sigma} g(t;\mu,\sigma) \ . \quad \textbf{(5)}$$

be the estimated signal. Then the mean squared error:

$$K^{-1} \sum_{t=1}^{K} [S(t) - \hat{S}(t)]^2 \quad ,$$

will be minimized by equating energy to cost and solving for **I** and **T**. Equation (4) has the form of equations (3) when each pair (μ,σ) is indexed by $i \in \{1,\ldots,N\}$, $N = ML$. In particular, the weight of the symmetric link between units $i \Leftrightarrow (\mu,\sigma)$ and $j \Leftrightarrow (\mu',\sigma')$ is:

$$T_{ij} = T_{(\mu,\sigma),(\mu',\sigma')} = -2 \sum_{t} g(t;\mu,\sigma) g(t;\mu',\sigma') \quad \textbf{(6a)}$$

and the biases are:

$$I_i = I_{(\mu,\sigma)} = 2 \sum_{t} S(t) g(t;\mu,\sigma) \ . \quad \textbf{(6b)}$$

The right sides of equations (6a) and (6b) cross-correlate the basis functions with each other and with the signal, respectively. Should it be required to have $V_i \in \{0,1\}$, the units can be McCulloch-Pitts neurons or the penalty term (2c), which does not affect **T**, can be added.

<u>Continuation</u>: Let the basis functions be of the form $\varphi(t;\mu,\sigma) = g(|t-\mu|;\sigma)$, i.e., $g(t;\mu,\sigma)$ is symmetric about $t = \mu$. Constrain the fit (5) to involve at most a single term for each time slice $\mu = 1,\ldots,K$:

$$\hat{S}(t) = \sum_{\mu=1}^{K} V_{\mu}\varphi(|t-\mu|;\sigma[\mu]) \quad ,$$

where $\sigma[1],\ldots,\sigma[K]$ is an assignment of a particular σ-parameter to each μ.

Solution: Let:

$$PENALTY(V) = \sum_{\mu}\sum_{\sigma}\sum_{\sigma'} CV_{\mu\sigma}V_{\mu\sigma'} \, , \quad \sigma \neq \sigma' \quad ,$$

with $C >> 0$. When each row vector $\mathbf{V} = (V_{\mu 1},\ldots,V_{\mu L})$ has a single nonzero component the penalty will vanish; but whenever two or more units in the same row are ON, a large penalty is incurred.

The McCulloch-Pitts neurons form a rectangular array with M rows and L columns. All units in the same row are linked by strongly inhibitory reciprocal synapses of strength -C. The addition of the penalty term modifies T_{ij} by the addition of:

$$T_{ij} = \begin{cases} -C & \text{if } i \text{ and } j \text{ share the same } \underline{\text{row}} \\ 0 & \text{otherwise} \end{cases}$$

to equation (6a). For C sufficiently large, this strong "lateral inhibition" guarantees that only valid solutions will be stable states.

Strong Lateral Inhibition and "Winner-take-all" Subnets

Optimization networks with strong lateral inhibition were introduced by Hopfield and Tank (1986c) with a physiological argument that might not bear repeating. The copious literature spun off by their work is unanimous in its focus on the computational implications. The functional role of strong lateral inhibition is to support data representations in which neurons of a given subnet interact through strong inhibitory links so as to evolve inexorably to stable states in which only one of the units in the subnet stays fully ON. This "winner-take-all" competition among the units of a subnet is the key to satisfying the constraints of many optimization problems. The (overlapping) subnets will usually correspond to rows or columns in a rectangular array of units.

Problem 5.4: Zora Thales (pronounced as in "sales") operates the most expensive matchmaking service in Gotham City. Her weekly task is to assess the compatibilities of single male and female clients in order to pair those best suited for holy matrimony. Once Zora has tabulated:

a_{mf} = the compatibility of Mr. [m] and Miss [f] > 0

for each m and f, the problem is to maximize the sum of the compatibilities subject to the constraint that no client be assigned more than one date, i.e., to maximize:

$$VALUE(V) = \sum_m \sum_f a_{mf} V_{mf}$$

subject to V_{mf}, $\Sigma_m V_{mf}$, and $\Sigma_f V_{mf} \in \{0,1\}$.

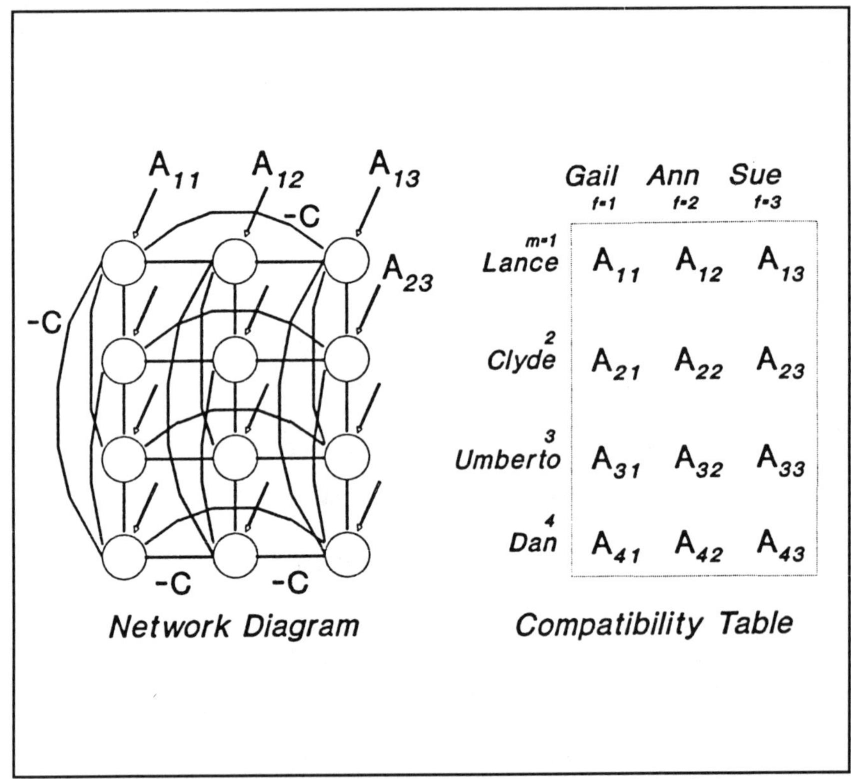

Figure 5.3 A collective computation network for solving an assignment problem.

Solution: Let the neurons be arrayed in an M x F rectangle where M and F are the numbers of male and female clients. Assume McCulloch-Pitts units. With reference to fig. 5.3, the units in each row are linked by strong inhibitory connections of strength -C; and likewise for the units in each column. The unit that indicates the pairing of Mr. [m] and Miss [f] has state V_{mf} (=1 if Mr. [m] and Miss [f] have a date and 0 if they don't) and receives a bias of a_{mf}. There are no other connections or inputs. The stable states of the network can be made to correspond to possible solutions by making C sufficiently large. For a stable state $\mathbf{V}^o$, the quadratic part of the computational energy vanishes. The energy of such a solution state is $E(\mathbf{V}^o) = -VALUE(\mathbf{V}^o)$. The turn-on of any unit, say V_{ij}, will then increase the energy by at least $E = C - a_{ij} >> 0$. The turn-off of a unit will increase the energy by a_{ij}. Thus $\mathbf{V}^o$ is stable.

Example 2, ctd.: The traveling salesman must stop in each city and visit no city more than once; and he cannot be in two places at one time. Indexing the pairs formed from the cities and stop numbers *serially* from 1 through $N = M^2$ in *any* order, these constraints are enforced by making:

$$T_{ij} = -C \text{ if i and j share the same row or column and } i \neq j, \quad \textbf{(7a)}$$

for $C >> 0$. (The reader is advised that a very simple case of this problem is worked out in the following example.) Again letting the row indicate the stop number, so that time advances from top to bottom, set:

$$T_{ij} = B - D(col[i],col[j]) \text{ if i and j are on } \textit{adjacent} \text{ rows but not in the same column} \quad \textbf{(7b)}$$

for some $B > 0$, where col[i] is the column in which the ith unit lies. A valid tour $\mathbf{V}^o = (V_1^o,\ldots,V_N^o)$ will have exactly one McCulloch-Pitts unit ON in each row and each column. Its energy will be:

$$E(V^o) = -\sum_{i<j} T_{ij}V_iV_j = D_T - MB$$

where (again) D_T is the tour length. Thus ENERGY = TOUR LENGTH - CONSTANT will be minimized in going from $\mathbf{V}(0)$ to $\mathbf{V}^o$.

Parameter identification in the TSP net is guided by two considerations. First the excitatory offset B must be at least as great as the greatest intercity distance $\sup_{i,j} D(i,j)$ in order for the links specified in equation (7b) to be positive. Second, the inhibitory strength C must at least equal the excitatory offset. If these conditions are satisfied, the solution states are stable. Turn-on of another unit (say the jth) adds:

$$\Delta E = 2C - 2B + D(j,.) + D(.,j)$$

to $E(\mathbf{V}^o)$. Turn-off of a unit adds:

$$\Delta E = 2B - D(j,.) - D(.,j).$$

Both ΔE's are positive when $B > \max[D(j,.),D(.,j)]$ where the maximum is taken over all D_{ij} and $C > B$. To clarify these concepts we illustrate this problem by a simple (because there are only a few cities) example.

Example 3: Let the salesman visit three cities: Annapolis (A), Baltimore (B), and Washington (W). The distances between town centers are

	A	B	W
A	0	40	45
B	40	0	50
W	45	50	0

Our salesman seeks to make a trip in which each city is visited exactly once. Let us introduce a 9x9 array of pairs consisting of a city and a stop number. Let C be a large positive number and there will be an inhibitory connection of strength -C between any two pairs that have a common city or a common stop number. The connection between any pair and itself is, of course, zero (i.e. the diagonal elements of the W matrix).

	A1	A2	A3	B1	B2	B3	W1	W2	W3
A1	0	-C	-C	-C			-C		
A2	-C	0	-C		-C			-C	
A3	-C	-C	0			-C			-C
B1	-C			0	-C	-C	-C		
B2		-C		-C	0	-C		-C	
B3			-C	-C	-C	0			-C
W1	-C			-C			0	-C	-C
W2		-C			-C		-C	0	-C
W3			-C			-C	-C	-C	0

B must be greater than any of the inter-city distances, so we chose B = 55. Then we fill in the entry for (A1, B2) by taking B -(the distance from Annapolis to Baltimore) = 55 - 40 = 15. Similarly for (B1, A2) and (B2, A1) and (A2, B1). (A1, W2), (A2, W1), (W1, A2) and (W2, A1) are all assigned a value 55 - 45 = 10. (A2, B3), (A3, B2), (B2, A3) and (B3, A2) are all assigned a value 55 - 40 = 15. And (A2, W3), (A3, W2), (W2, A3) and (W3, A2) are all 55 - 45 = 10.

	A1	A2	A3	B1	B2	B3	W1	W2	W3
A1	0	-C	-C	-C	15		-C	10	
A2	-C	0	-C	15	-C	15	10	-C	10
A3	-C	-C	0		15	-C		10	-C
B1	-C	15		0	-C	-C	-C		
B2	15	-C	15	-C	0	-C		-C	
B3		15	-C	-C	-C	0			-C
W1	-C	10		-C			0	-C	-C
W2	10	-C	10		-C		-C	0	-C
W3		10	-C			-C	-C	-C	0

If Annapolis (A) is number 1 and Baltimore (B) is number 3 the trip must be completed by a return journey to Annapolis (A). So, we assign 55 - 40 = 15 to (A1, B3), (A3, B1), (B1, A3) and (B3, A1). Finally, we assign 55 - 45 = 10 to (A3, W1), (A1, W3), (W1, A3) and (W3, A1). This gives us the matrix below.

The reader is advised that this problem is useful only for clarifying the concepts of the preceding problem. The total (Euclidean) distance traversed will be the perimeter of a triangle, regardless of the order in which the cities are visited. Hence, every solution is an optimal solution. To the reader who thinks this is too condescending, we rejoin that, in the years following Hopfield's initial description of the TSP neural circuit many researchers seemed to believe that the Hopfield net was "learning" to adjust its weights instead of cranking out solutions with the T_{ij}'s all figured out a priori.

	A1	A2	A3	B1	B2	B3	W1	W2	W3
A1	0	-C	-C	-C	15	15	-C	10	10
A2	-C	0	-C	15	-C	15	10	-C	10
A3	-C	-C	0	15	15	-C		10	-C
B1	-C	15	15	0	-C	-C	-C	5	5
B2	15	-C	15	-C	0	-C	5	-C	5
B3	15	15	-C	-C	-C	0	5	5	-C
W1	-C	10	10	-C	5	5	0	-C	-C
W2	10	-C	10	5	-C	5	-C	0	-C
W3	10	10	-C	5	5	-C	-C	-C	0

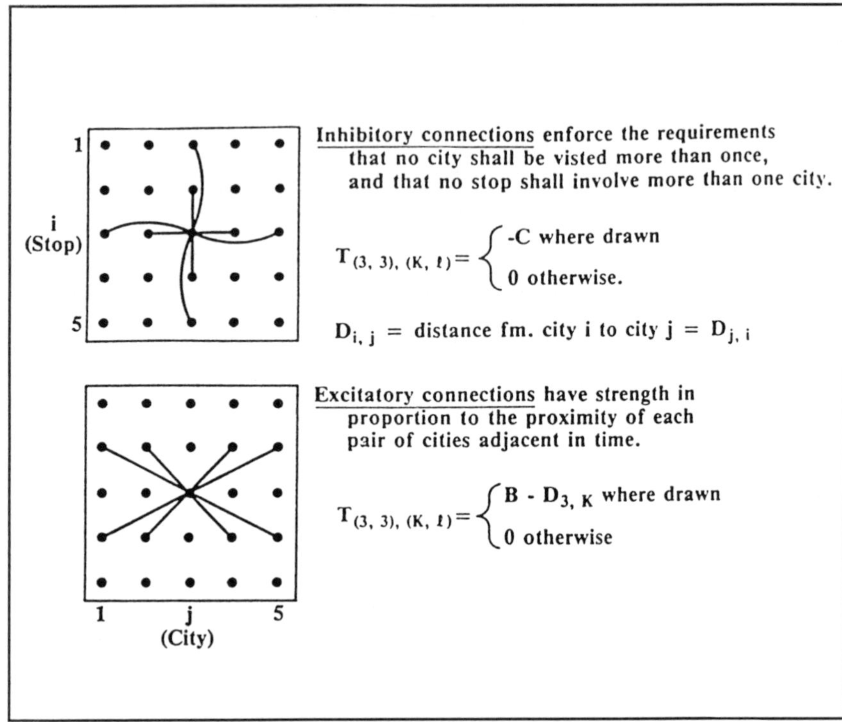

1
i
(Stop)
5
Inhibitory connections enforce the requirements
that no city shall be visted more than once,
and that no stop shall involve more than one city.
T (3, 3), (K, ℓ) = -C where drawn
0 otherwise.
D i, j = distance fm. city i to city j = D j, i
Excitatory connections have strength in
proportion to the proximity of each
pair of cities adjacent in time.
T (3, 3), (K, ℓ) = B - D 3, K where drawn
0 otherwise
1
j
5
(City)

Figure 5.4 Fixing the Weights to Solve the five-City TSP

Problem 5.4: Let V(i,j) = 1 if the jth city is visited at stop i-1 for i and j = 1,...,M; and V(i,j) = 0 otherwise. Describe an algorithm for creating a weight matrix and finding solutions to the M-city TSP as an M x M permutation matrix **V**.

Solution: The following five steps begin with the assumption of a Euclidean TSP:

(1) Input the coordinates of each city and compute the M(M-1)/2 intercity distances.

(2) Dimension an M x M x M x M array T(i,j,k,m) that will describe the connectivity among McCulloch-Pitts neurons that are indexed by number pairs. For instance, T(1,1,M,M) is the strength of the symmetric link between unit (1,1), in the upper left corner of the V-array, and unit (M,M) in the lower right. Let:

$$T(i,j,k,m) = -C \quad \text{if } i=k \text{ or } j=m \qquad \textbf{(8a)}$$

so that, for $C >> 0$, the turn-on of more than one unit per row or column will be strongly inhibited. Also let:

$$T(i,j,k,m) = B - D(j,m) \quad \text{if } i = k \pm 1 \bmod M \text{ and } j = m, \qquad \textbf{(8b)}$$

so that, for $B > D(.,.)$, the strength of the excitatory link between two units in adjacent rows decreases for increasing distance between cities j and m. Symmetrize the array so that:

$$T(k,m,i,j) = T(i,j,k,m).$$

Finally insist that $T(i,j,i,j) = 0$. Figs. 5.5 and 5.6 are intended to clarify this step. In particular, fig. 5.6 shows the wrap-around provisions that link the city of origin to the Mth stop in the tour.

(3) Dimension an MxM square array V(i,j) and set its elements to zero. V(i,j) will subsequently become the indicator of whether the jth city is visited at stop i-1, stop zero being the origin of the tour. (Alternatively, let the V(i,j) be independent Bernoulli random variables.)

(4) Select a pair (i,j) at random and recompute V(i,j) using the rule:

$$V(i,j) = \mathcal{H}\left[\sum_{m=1}^{M}\sum_{k=1}^{M} T(i,j,k,m) + I(i,j)\right] \qquad \textbf{(9)}$$

where the biases **I** are zero except for $I(0,j[0]) >> C$ to clamp ON the unit representing the city of origin.

(5) Iterate equation (9) until **V** no longer changes. Presumably the stable state will be reached after several sweeps.

Fig. 5.6 suggests that the simplest cases of Euclidean TSP have optimal solutions represented by equilateral polygons with M sides. Each tour in the figure has a counterpart in which the directional arrows are reversed.

Figure 5.7 shows some computed results for the case of the equilateral pentagon. The parameters B and C are stepped through their ranges a unit at a time. Each step is followed by a numerical test in which a random initial state is chosen and equation (9) is iterated until convergence with $\mathbf{I} = 0$ (no city-of-

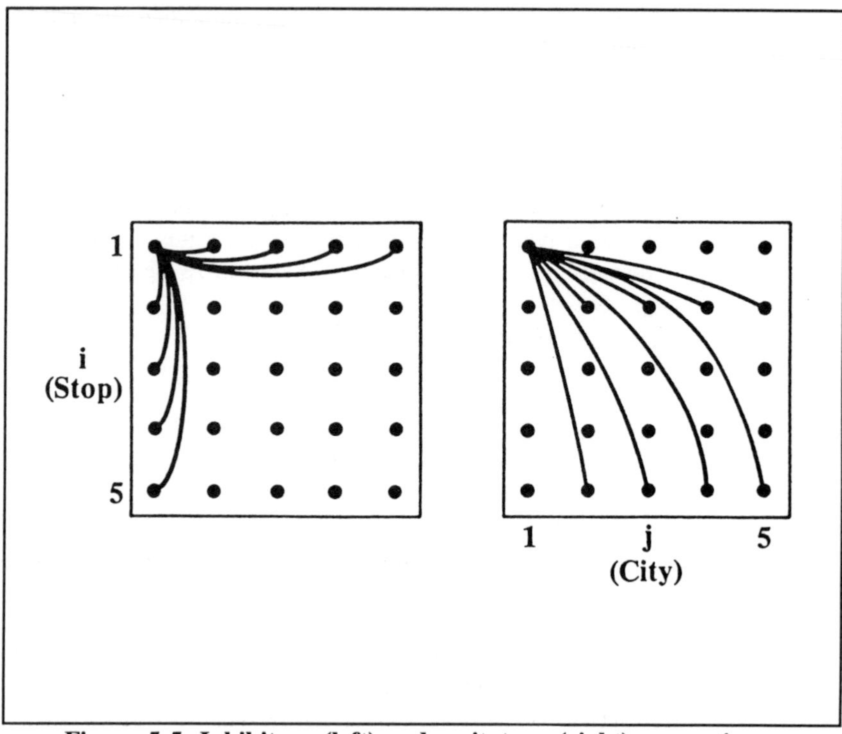

Figure 5.5 Inhibitory (left) and excitatory (right) connections.

origin specified). Then for each pair (B,C) we obtain a stable state $\mathbf{V}^{\circ}$ with activity:

$$A(V^{o};B,C) = \sum_{i=1}^{M} \sum_{j=1}^{M} V_{ij}^{o}$$

which counts the number of units ON. These results are consistent with earlier remarks on parameter identification. In general, we use:

$$B = \sup_{i<j} D(i,j) \quad \text{and} \quad C = 2B.$$

Suboptimal Solutions

Figure 5.7 shows solution quality versus frequency in forty independent trials of the (unit pentagon) TSP. Twenty trials started from $\mathbf{V}(0) = \mathbf{0}$ with a strong bias applied to the origin unit. Twenty trials started from $\mathbf{V}(0)$ consisting of random bits with $\mathbf{I} = \mathbf{0}$. The resulting tours are of four kinds, discounting their permutations, with tour lengths ranging from 5.875 units (the optimum) to 8.1 (the pentagram/star). The figure indicates that the optimum was attained

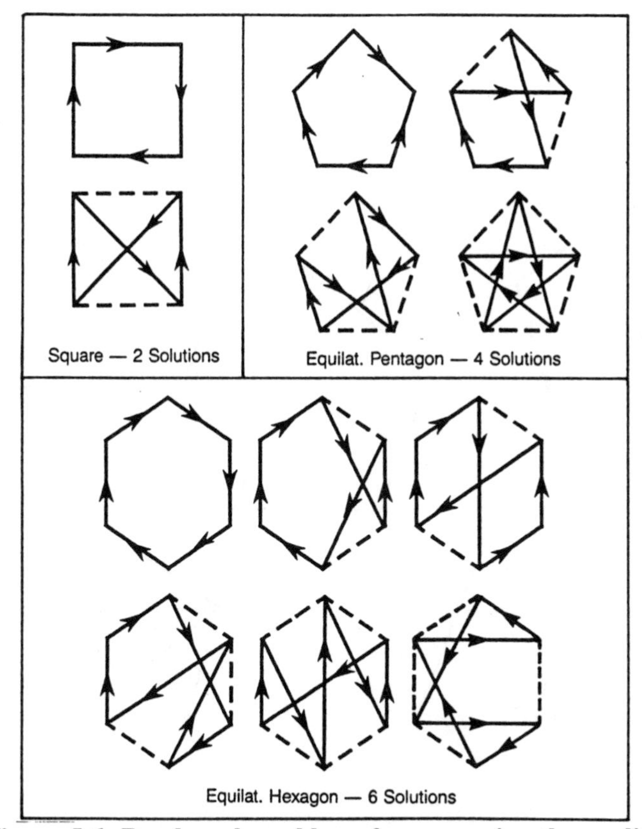

Figure 5.6 Benchmark problems for measuring the quality of neural net (TSP) solutions.

1 (the pentagram/star). The figure indicates that the optimum was attained ur times out of forty.

Figure 5.8 is a normalized, accumulated histogram of the tour length ost) in forty independent trials of a non-Euclidean five-city TSP defined by the ble below it. Each trial began with $\mathbf{V}(0)=\mathbf{0}$ and a strong bias to the unit in e upper-left corner of the neuron array. The data suggests that about half the lutions are good and that the other half are essentially quite bad.

Figure 5.9 conveys a similar lesson. The plotted points mark some of e percentiles of the accumulated distribution of cost in randomly-generated, n-Euclidean, five-city TSPs. The straight line "fit" ignores the evident S-ape of the distribution function in question; but it confirms the suspicion that lution quality is almost uniformly distributed between the best and the worst.

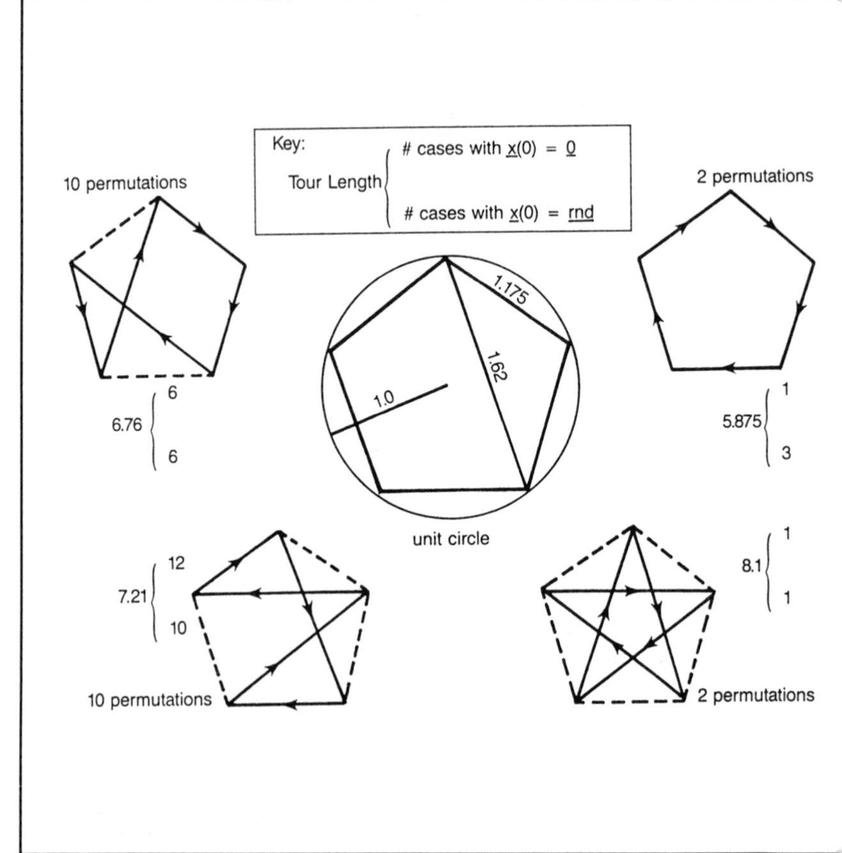

Figure 5.7 Solution quality and frequency (x20) for TSP-5.

These disappointing results are rationalized by fig. 5.10 in which t stable state evolves in a manner consistent with the constraints imposed equations (7a) and (8a) and propelled by the dynamical equation (9) of t stochastic model toward stable states representing valid tours. While t constraints are satisfied, the network exhibits no preference for better solution Indeed, when we look beyond the overt constraints that establish valid solution the detailed evolution of the network state is controlled by the random numb generator of the dynamical process; and it is not surprising that the resu evince a perfect ignorance regarding solution quality. Figure 5.11 shows th resort to the deterministic dynamics of the synchronous update model will n overcome this difficulty.

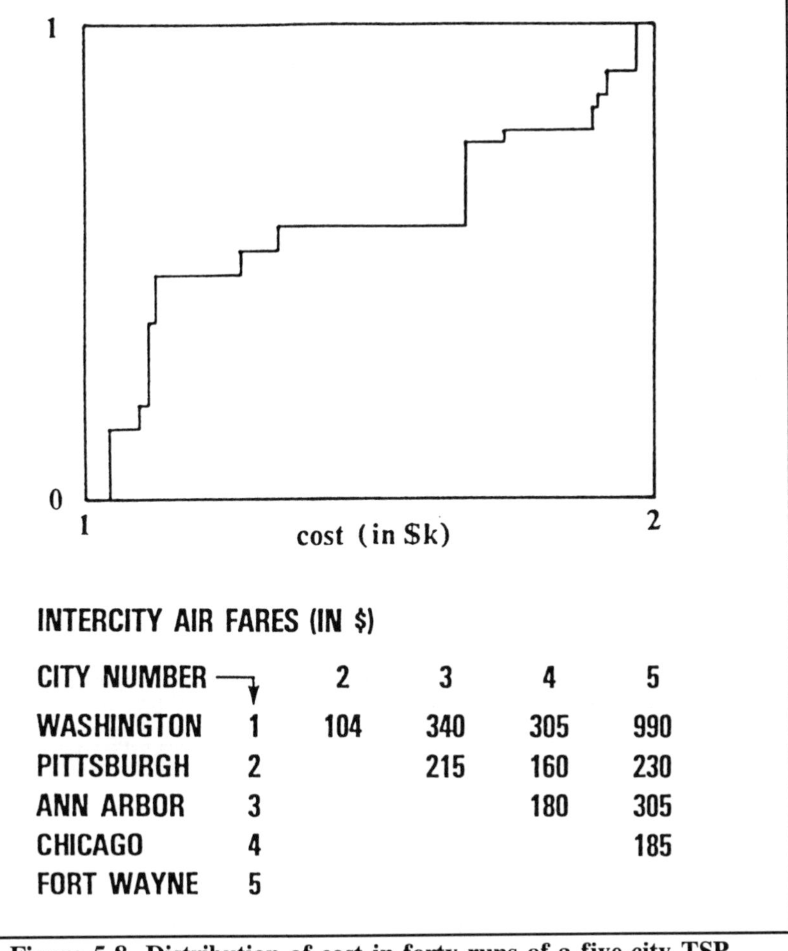

INTERCITY AIR FARES (IN $)

CITY NUMBER		2	3	4	5
WASHINGTON	1	104	340	305	990
PITTSBURGH	2		215	160	230
ANN ARBOR	3			180	305
CHICAGO	4				185
FORT WAYNE	5				

Figure 5.8 Distribution of cost in forty runs of a five-city TSP.

Problem 5.5 (Peak Picker): Devise a network of six McCulloch-Pitts neurons to find the largest of six positive numbers. Evaluate its performance assuming the stochastic dynamics.

Solution: The sketch for problem 3.12 will suffice. Each unit inhibits all the others and receives a bias equal to one of the numbers to be selected. Starting from $\mathbf{V}(0) = \mathbf{0}$, the network converges in one time step to any one of the six stable states with *the same probability*. The depths of the energy valleys

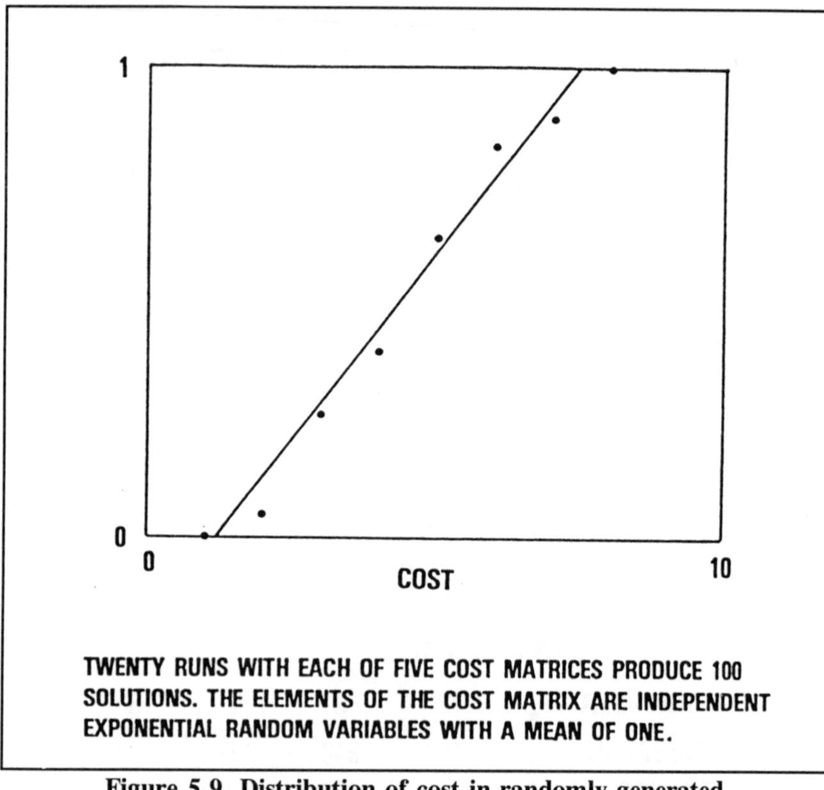

Figure 5.9 Distribution of cost in randomly generated non-Euclidean, five-City TSPs.

at the stable states are proportional to solution quality; but the *basins of attraction*, as defined in chapter 3, all have the same size.

Another illustration: In a resource allocation problem like problem 5.2, let $N = 10$, $a_i = i$, and $C_i = 10-i+1$, with $b << 55$. The optimum solution is obviously $V_i = \mathcal{H}(i - i^*)$, i^* being the largest integer such that $1 + 2 + 3 \ldots i^* \leq b$. The table below shows some results computed with $b = 6$ for each of three values of d. The number of optimal solutions found by the network was fourteen out of sixty total trials of this (very easy) problem.

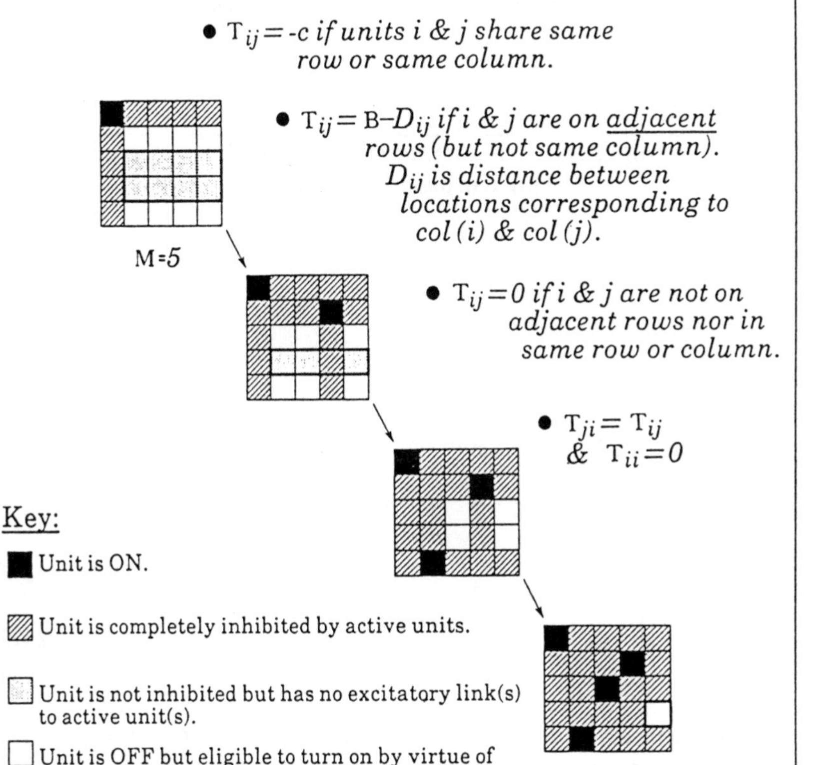

Evolution of State in a TSP Network with Glauber Dynamics
Let there be M stops, corresponding to the rows, with the point-of-origin specified by the first (top) row, etc. The connectivity is described by
• T_{ij} = -c if units i & j share same row or same column.
• T_{ij} = B–D_{ij} if i & j are on adjacent rows (but not same column). D_{ij} is distance between locations corresponding to col (i) & col (j).
• T_{ij} = 0 if i & j are not on adjacent rows nor in same row or column.
• T_{ji} = T_{ij} & T_{ii} = 0
M=5
Key:
Unit is ON.
Unit is completely inhibited by active units.
Unit is not inhibited but has no excitatory link(s) to active unit(s).
Unit is OFF but eligible to turn on by virtue of excitatory links to active unit(s).

Figure 5.10 Evolution of state in a TSP network with Glauber dynamics.

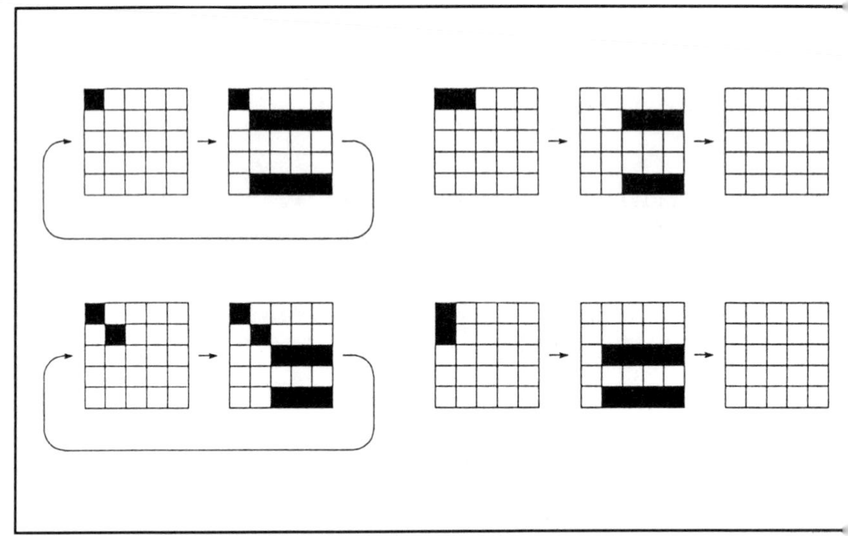

Figure 5.11 Evolution of state in a TSP network with synchronous dynamics.

COST CRITERION = 6 units. NO. CORRECT

VALUE	1	2	3	4	5	6	7	8	9	10	NO. CORRECT		averages
cost	10	9	8	7	6	5	4	3	2	1			
d = 3	0	0	3	0	2	7	2	6	8	17	6	6.5	18.2
d = 2	0	0	2	1	0	7	6	5	16	18	4	6.8	22.8
d = 1	0	0	0	1	2	9	4	5	18	20	4	7.5	24.9

In summary, our procedures for shaping the energy landscape have been effective in creating deeper valleys at better solutions, but totally ineffective in so far as the longitudinal extent of the basin of attraction is concerned. There is no obvious proportionality between depth and longitudinal extent in the energy landscape — or else our stochastic model is incapable of illuminating it. Initial states evolve to stable states with probabilities uninfluenced by the relative depths. Hopfield and Tank (1986c) had warned us against this discovery:

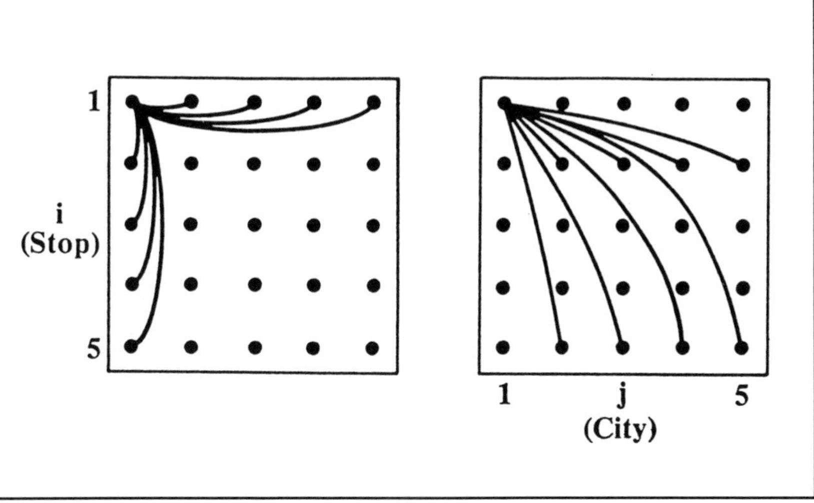

Figure 5.12 Weights to Solve the five-city TSP (continued.) Inhibitory (left) and excitatory (right) connections showing wrap-around of the latter.

> The analog nature of the neural response in this problem [is] essential to its computational effectiveness. This use of a continuous variable between true and false is similar to the theory of fuzzy sets and to the use of evidence voting for the truth of competing propositions in Bayesian inference and connectionist modeling in cognitive psychology. Two-state neurons do not capture this computational feature.

aving disregarded this warning, we observed the collective computational ilure of the McCulloch-Pitts neurons in the optimization network. Moreover, e most evident cause of this failure is the randomness of the dynamics.

Kahng (1989) has followed Wilson and Pawley (1988) in trying to onfirm the preference for good (suboptimal) solutions found by Hopfield and ank (1985) using continuous dynamics and graded neurons. Kahng, running n city TSPs with parameters prescribed by Hopfield and Tank, found that our quality is good when a valid tour is found, but unfortunately only about 5% of the outputs are valid tours." In Kahng's words,

The obvious problem is that the terms of the energy function separately attempt to enforce structure (ie, a permutation matrix output) and cost [tour length] minimization. When structure is enforced, output tours are of poorer quality. When we attempt to minimize cost, invalid outputs result. A host of papers are essentially devoted to scaling heuristics designed to make the Hopfield-Tank method yield correct tours. It may be asked whether "fixing" the Hopfield-Tank approach to the TSP is worthwhile.

We will sidestep this controversy by keeping our binary units and retaining th stochastic model — and by adding a new dimension of randomness to th dynamics. In the terminology of Hinton and Sejnowski, we will use **noise** t escape from shallower local minima and enable the network state to settle int deeper energy valleys representing better solutions. The result, in the nex chapter, will be a "Boltzmann machine" in which the network state wander ergodically through state space, visiting the states with relative frequencie determined by their energies. Recalling chapter 4, this strategy should b entirely predictable to the statistical physicist. This midway point is a watershe in the present text. Up to now, we have studied collective computation network using the stochastic (Glauber) dynamics at **zero temperature**. Subsequently w raise (and lower) the effective temperature of the network in search of bette solutions to the given problem.

6. Markov Chains and Markov Fields in Collective Computation Networks

Belief in the possibility of "computers that think like humans" is older an digital computing. Von Neumann (1954) suggested that brain theory might e developed along the lines of statistical physics, as if large numbers of neurons ace paths in a phase space through their collective action. The coordinates of point in this space, corresponding to a cognitive state, would specify the states f the neurons. Then a distribution in this space, assigning probabilities to oints conditioned on given stimuli, could represent an intelligent relation to ome environment.

By disclosing a learning algorithm for "Boltzmann machines", Ackley, Iinton and Sejnowski (1985) drew attention to an idea sketched earlier by Iinton and Sejnowski (1983) in one of the first constructive reactions to the Iopfield (1982) model. At the risk of seeing only what we want to see, we ummarize their idea as follows:

(a) In symmetrically interconnected networks of McCulloch-Pitts neurons, the individual units act so as to collectively minimize the global energy:

$$H(\mathbf{x}) = -\tfrac{1}{2} \sum_j \sum_i W_{ij}\, x_i\, x_j \; - \sum_i (z_i - \theta_i) x_i \; , \qquad (1)$$

where $\mathbf{x}$ is the network state vector, $\mathbf{W}$ is the symmetric NxN weight matrix, $\mathbf{z}$ is a vector of externally-applied inputs (the biases), and θ_i is the threshold of the ith unit.

(b) The *energy gap* associated with the ith unit is:

$$\Delta H_i = H(x_1,\ldots,x_i{=}0,\ldots,x_N) - H(x_1,\ldots,x_i{=}1,\ldots,x_N) \qquad (2)$$

$$= \sum_j W_{ij}\, x_j + z_i - \theta_i$$

The network "suffers from the standard weakness of gradient descent methods: It gets stuck in *local* minima that are not globally optimal. This is the inevitable consequence of allowing only jumps to states of lower energy." (Hinton and Sejnowski, 1983).

(c) "If jumps to higher energy states occasionally occur, it is possible to break out of local minima." Hinton and Sejnowski adapted the Metropolis (1953) algorithm and the simulated annealing approach of Kirkpatrick et al. (1983). "If the energy gap between the true [ON] and false [OFF] states of the kth unit is $+\Delta H_k$ then, regardless of the previous state, set $x_k = 1$ with probability:

$$p(\Delta H_k) = \frac{1}{1 + \exp(-\Delta H_k / T)} \tag{3}$$

where T is a parameter that acts like *temperature* (fig. 6.1)."

(d) This stochastic algorithm "ensures that in *thermal equilibrium* the relative probability of two global states...follows a Boltzmann distribution:"

$$P(\mathbf{x})/P(\mathbf{x}') = e^{-[H(\mathbf{x}) - H(\mathbf{x}')] / T} \quad . \tag{4}$$

Hinton and Sejnowski went on to develop a formula for the weights based on a Bayesian statistical argument. (This formula will be rederived and used as an alternative to the outer product rule in chapter 8 when we return to content-addressable memory.) They also showed how the weights could be trained incrementally. This method, elaborated by Ackley, Hinton, and Sejnowski (1985), and Sejnowski, Kienker, and Hinton (1986), is outside the scope of the present text. This chapter and the next are merely intended to clarify the mathematical principles on which the last three equations are based.

Let $X_i(t) \in \{0,1\}$ be the state of the ith unit, $i = 1,\ldots,N$, at time $t = 0,1,2,\ldots$ Use $\mathbf{X}(t) = (X_1(t),\ldots,X_N(t))$ for the network state vector, the upper case symbolizing a *random* variable, so that $\mathbf{x}$ is a particular realization of $\mathbf{X}$; but retain $\mathbf{W}$ for the symmetric N x N weight matrix with all zero diagonal elements. Now $\{K(t),\ t = 0,1,2,\ldots\}$ is a sequence of independent random integers uniformly distributed among the first N integers, which index the neurons. $\mathbf{\Theta} = (\theta_1,\ldots,\theta_N)$ is the vector of time-invariant thresholds (if any). The missing link that unifies equations (1) through (4) is the dynamical equation:

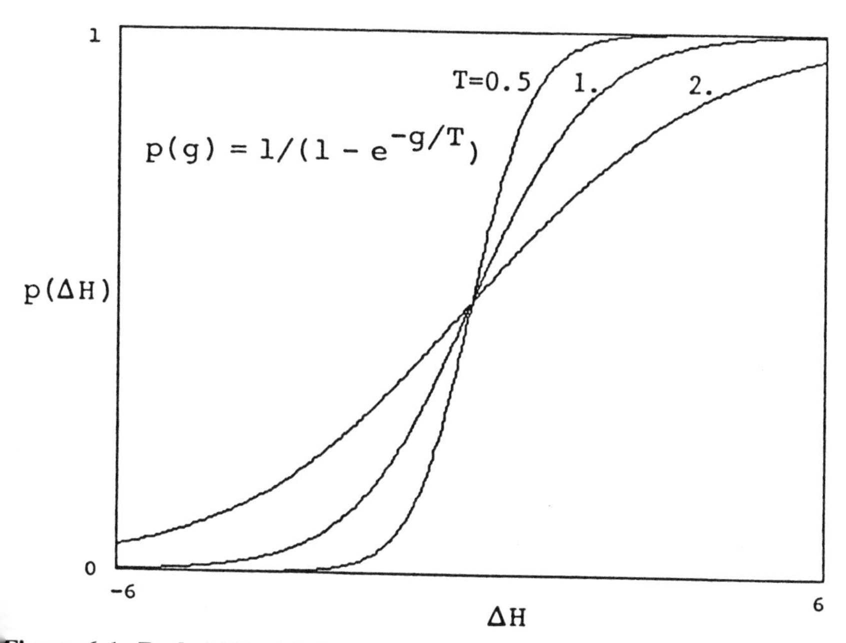

Figure 6.1 Probability (p) that a unit turns ON to close the energy gap (ΔH) for several values of T.

$$X_i\ (t+1) = \begin{cases} \mathcal{H}\left[\sum_{j=1}^{N} W_{ij}\ X_j(t) + z_i\ (t) - \theta_i\right] & \text{if } i=K(t) \\ X_i\ (t) & \text{otherwise} \end{cases} \qquad \textbf{(5)}$$

which generates the sequence $\{\mathbf{X}(t),\ t \geq 0\}$ from initial state $\mathbf{X}(0)$. The main points are contained in two theorems. Before proceeding to the theorems, we introduce two definitions:

Convergence in probability means that by taking t sufficiently large we can achieve an arbitrarily high probability (≤ 1) of having $\mathbf{X}(t) \in \{\mathbf{x}^o\}$.

Convergence in distribution means that, for t sufficiently large, the probability of having $\mathbf{X}(t) = \mathbf{x}$ is $P(\mathbf{x})$ regardless of the initial state.

Theorem 6.1 (Hopfield): If $Z_i(t) = z_i$ is constant then the sequence $\{\mathbf{X}(t$ $t \geq 0\}$ of network states converges *in probability* to an $\mathbf{x}^o$ that is one of the loc minima of equation (1).

Proof: The sequence is a Markov chain which is absorbed with probabili one by an irreducible persistent set as noted in Chapter 3. When **W** satisfies th stated conditions, these irreducible sets have cardinality one and coincide wi the local minima of the energy function.

Theorem 6.2 (Hinton and Sejnowski): If $\{Z_i(t), t=0,1,2,\ldots\}$ is a sequen of independent, identically distributed (i.i.d.) *logistic* random variables wi mean μ_i and common scale parameter T, i.e., if:

$$Pr[Z_i(t) \leq z] = \frac{1}{1 + \exp[-(z-\mu_i)/T\ ]} \tag{6}$$

for every i and t, then $\{\mathbf{X}(t), t \geq 0\}$ converges *in distribution* to th Gibbs/Boltzmann law. This distribution is:

$$P(\mathbf{x}) = \frac{\exp[-H(\mathbf{x})/T\ ]}{\sum_{\mathbf{x}'} \exp[-H(\mathbf{x}')/T\ ]} \tag{7}$$

where $\mathbf{x}'$ ranges through all 2^N states and H is the same energy function a equation (1), but with μ_i substituted for z_i. Recall (chapter 4) that th denominator on the right in (7) is the *partition function.*

The meaning of theorem 2 might warrant some clarification before th question of proof is addressed.

This chapter is no place for a detailed review of the development o statistical mechanics and its rationalization of the laws of thermodynamics fro the viewpoint of molecular theory. It should suffice to recall that Gibbs (190 considered constitutionally identical multiparticle systems each defined by a poir in phase space, the (Hamiltonian) coordinates of which specify the position an momentum of each constituent particle. To each point in phase space, an hence to each possible system configuration, there corresponds an energy H an a probability P. Gibbs showed that the canonical relation between ln(P) and H was linear, as $\ln(P) = (\text{const.} - H)\beta$; and $1/\beta$ turns out to be proportional to th temperature.

Probability has two equivalent meanings in this context by virtue of the rgodic property. If the state **X** of a single system, initially **X**(0), is observed times t=1,2,..., then:

$$\lim_{t\to\infty} \Pr[\mathbf{X}(t) = \mathbf{x}] = P(\mathbf{x}).$$

he meaning of this statement can be interpreted in light of the frequency efinition of probability: If m(**x**) is the number of times at which **X**(t) = **x** up t = M, then the relative frequency of **x** converges (in probability) to P(**x**):

$$\lim_{M\to\infty} m(\mathbf{x})/M = P(\mathbf{x}).$$

Ioreover, if we have an ensemble of M constitutionally identical but physically ndependent systems, and we survey them all at a single instant in time, then the st limit is the same when m(**x**) is interpreted as the number of systems stantaneously in state **x**.

Problem 6.1: Prove theorem 2 for the special case N = 1.

Solution: Let the subscript i = 1 be dropped so that:

$$\Pr[\mathbf{X}(t+1) = 1] = \Pr[Z(t) > \theta]$$

ollows directly from the dynamical equation (5). (Remember $W_{11} = 0$ and her W's do not exist for N = 1.) But:

$$\Pr[Z(t) > \theta] = 1 - \Pr[Z(t) \leq \theta]$$

$$= 1 - 1/[1 + e^{-(\mu-\theta)/T}]$$

ith reference to equation (6). Rearrange the last expression to obtain:

$$\Pr[\mathbf{X}(t + 1) = 1] = e^{(\mu-\theta)/T}/(1 + e^{(\mu-\theta)/T}) \, . \qquad \textbf{(7a)}$$

ince H(1) = θ - μ and H(0) = 0, the partition function is the denominator of e right hand side and (7a) is indeed an instance of equation (7).

If the single neuron in question is one of many and the rest are *amped*, the total input is U = Y + Z where $Y = Y_i = \Sigma_j W_{ij} x_j$ is the net input d the threshold is zero. Let the bias Z(t) be generated by a *noise source*. The oise is spectrally white (flat) since {Z(t), $t \geq 0$} consists of independent (and erefore uncorrelated) random variables. The noise amplitude probability

density function is f(z), a symmetric (ie: even; f(z) = f(-z)), bell-shaped curve
The (cumulative) distribution function is:

$$F(z) = Pr[\ Z(t) \leq z] = \int_{-\infty}^{z} f(x)\ dx \quad .$$

The evenness of f(.) is both necessary and sufficient for F(z) = 1 - F(-z).

Fig. 6.2, in which V is the neuron state, shows that, under thes conditions, the probability that the unit turns ON to a net input Y=y is simply

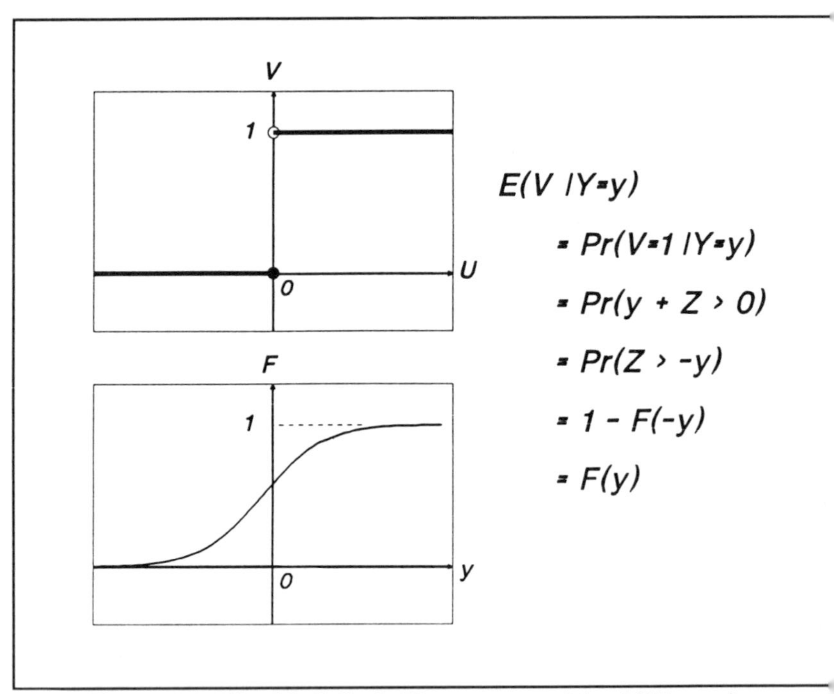

Figure 6.2 The McCulloch-Pitts neuron with additive noise.

F(y). Moreover, since V is 0-1 binary, its expected value conditioned on havin Y=y, written E(V|Y=y), is the same as the turn-on probability. Wit reference to equation (2), the net input in this case is the same as the "energ gap"; and if the situation were redefined with nonzero μ and θ, the turn-o probability would still be F(ΔH) in terms of the gap. To be consistent wit equation (3) we require $F(y) = 1/(1 + e^{-y/T})$. This is the logistic distributio function with zero mean and scale parameter T. Figure 6.3 shows the normal (N1), logistic (N2), and Cauchy (N3) density functions:

$$\text{(N1)}\ \exp(-z^2/2)/\sqrt{2\pi},$$

$$\text{(N2)}\ \exp(-z)/\ [1+\exp(-z)]^2,$$

and

$$\text{(N3)}\ 1/[\pi(1+x^2)],$$

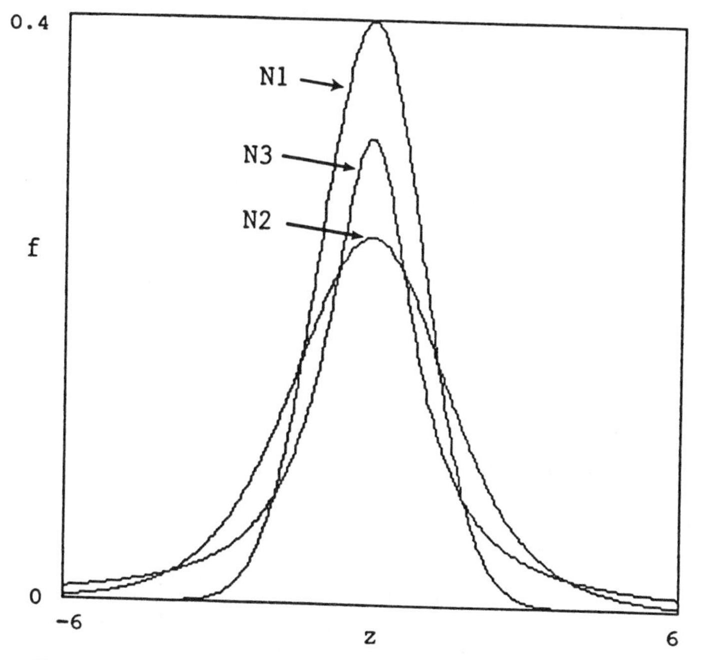

Figure 6.3 Three probability densities: the normal (N1), the logistic (N2), and the Cauchy (N3), all with unit scale parameter.

espectively, corresponding to the distribution functions of fig. 6.4 with scale parameter $\sigma=1$. Note that σ is the root variance (or standard deviation) of the normal distribution; but the variance of the logistic random variable is:

$$\int_{-\infty}^{+\infty} z^2 e^{-z/T}(1+e^{-z/T})^{-2} dz/T = (\pi T)^2/3 \approx (1.81T)^2 \quad .$$

Normal

$$\text{N1:} \quad F(z) = \int_{-\infty}^{z} \frac{1}{\sqrt{2\pi}\,\sigma} e^{-x^2/2\sigma^2} dx$$

Logistic

$$\text{N2:} \quad F(z) = 1/(1 + e^{-z/\sigma})$$

Cauchy

$$\text{N3:} \quad F(z) = \frac{1}{2} + \frac{1}{\pi} \text{Arctan}\left(\frac{z}{\sigma}\right)$$

Figure 6.4 Alternatives for F(z). Note that in all cases: F(-z) = 1 - F(z).

Problem 6.2: Let X be the state of a McCulloch-Pitts neuron and let $F(z) = 1 - F(-z)$ be an S-shaped distribution function (d.f.) with an underlying bell-shaped probability *density* function. Show that the expected value of X is the same when conditioned on any of the following sets of assumptions:

(a) For net input y, fixed threshold θ, and constant bias z, the unit turns ON with probability $F(y+z-\theta)$.

(b) For net input y, fixed threshold, and randomly varying bias:

$$Z(t) = z + \Xi(t),$$

$$\Pr[\Xi(t) \leq \xi] = F(\xi), \qquad (*)$$

the unit assumes state $\mathcal{H}(y+Z(t)-\theta)$.

(c) For net input y, fixed bias z, and randomly varying threshold the unit assumes state $\mathcal{H}(y+z-\Theta(t))$. Assume the same distribution function (*).

The solutions should be evident from the foregoing remarks. The interpretation is just as interesting. The stochastic neuron (a) introduced by Hinton and Sejnowski is statistically the same as a deterministic neuron that is (b) buffeted by externally applied noise or (c) characterized by a randomly fluctuating threshold.

The origin of the noise in the additive noise model (b) might not be "external" at all: It could be a simplified representation of the synaptic noise due to randomness in the amount of neuro-transmitter released by an axon terminal or to the action of "stray" neurotransmitters that have not been completely scavenged by the enzymes that serve that purpose.

Problem 6.3: Let the excitation (or inhibition) of unit i due to unit j be $\widetilde{W}_{ij} x_j$, where $\widetilde{W}_{ij} = W_{ij} + \Delta W_{ij}$, and the elements of $\mathbf{\Delta W}$ are independent identically distributed (i.i.d.) random variables with zero mean and unit variance. $\mathbf{W}$ is symmetric with zero diagonal as before. In the absence of thresholds or biases, the total input to unit i is $\Sigma_j W_{ij} x_j + \Xi$. Assuming that ΔW_{ij} is sufficiently small so that the values of $\mathbf{x}$ may be considered to be independent of the ΔW_{ij}, find the mean and variance of Ξ.

Solution: The mean is zero, since:

$$E\left(\sum_{j=1}^{N} \Delta W_{ij} x_j\right) = \sum_j \left(E\Delta W_{ij} \right) E x_j = 0 \ .$$

Recall $E(\Delta W_{ij}) = 0$.) The variance is equal to the mean *activity*:

$$E\left(\sum_j \Delta W_{ij} x_j\right)^2 = E\left[\sum_j (\Delta W_{ij})^2 x_j^2 + 2\sum_j \sum_k \Delta W_{ij} \Delta W_{ik} x_j x_k\right].$$

ince the ΔW's have zero mean and are independent, $E(\Delta W_{ij} \Delta W_{ik}) = 0$. Ioting $x_j^2 = x_j$ ($\in \{0,1\}$), we have:

$$E\left(\sum_j \Delta W_{ij}\, x_j\right)^2 = \sum_j E\left(\Delta W_{ij}\right)^2 E x_j = E\left(\sum_j x_j\right) ,$$

ith the last step due to $E(\Delta W_{ij})^2 = 1$.

It would seem that, for large N, the assumptions of the last problem lead by way of a central limit theorem to a normal distribution of Ξ. Then the modeling of synaptic noise as a random bias implies normal (Gaussian) noise with variance modulated by the activity of the net. The Gaussian noise assumption, however, does not lead exactly to the Boltzmann machine, as problem 6.1 and subsequent exercises will demonstrate.

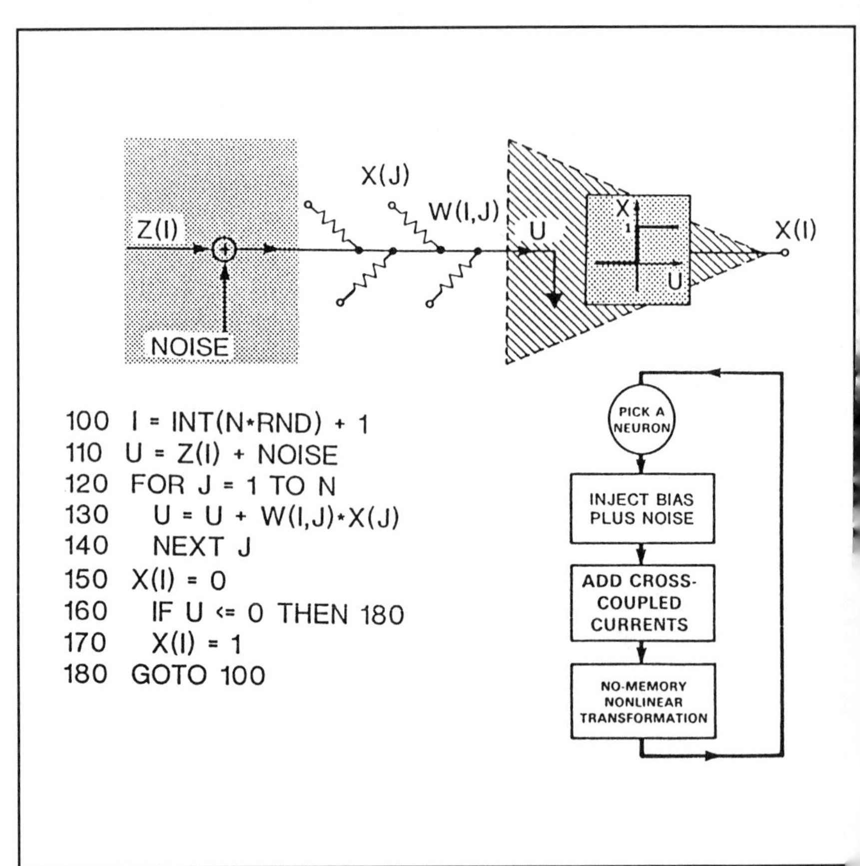

Figure 6.5 Programming the Boltzmann machine.

The additive noise model is easy to program. Figure 6.5 offers thre versions of the dynamics, the BASIC program lines and the flow diagram bein consistent with equation (5) after abolishing the threshold (which can always b subtracted from the fixed part of the bias). Line 100 picks a neuron [i = K(t and 110-140 sum up the total input. Missing from the figure is line:

```
105 NOISE = TEMP*LOG(1/RND - 1)
```

which generates logistic random variables with a scale factor equal to the temperature. (Note that LOG (in BASIC) takes the natural logarithm.)

In general, the programmer generates random variables with distribution function F(z) from computer-generated, uniform (unit) random variables R (= RND), $0 \leq R \leq 1$. Specifically, the transformation $Z = F^{-1}(R)$ is used, were F^{-1} is the inverse distribution function: $F[F^{-1}(R)] = R$.

Problem 6.4: Find the inverse of the logistic distribution function.

Solution: $R = F(Z) = (1 + e^{-Z/T})^{-1} \Rightarrow 1/R - 1 = \exp(-Z/T) \Rightarrow$ $Z = -T*\ln(1/R - 1)$. If the mean is $\mu \neq 0$, then $Z = \mu + T \ln(1/R-1)$. A fairly readable discussion of this procedure for other distributions is contained in a paper by Baran (1988).

The BASIC program in fig. 6.5 accomplishes no intelligent purpose in its endless DO loop; but if it were modified so that the number of times m(**x**) that state **x** is visited in the course of M iterations is tallied, the quotients m(**x**)/M would all approach the P(**x**) in theorem 2.

Transition Matrix

With the same assumptions as Theorem 2, $\{\mathbf{X}(t), t \geq 0\}$ is a homogeneous, irreducible, aperiodic Markov chain (HIAMC) on all 2^N states. The Markov chain is *homogeneous* because the parameters **W**, **Θ**, and EZ are constant. That it is *not periodic* seems plausible since there is nothing in the dynamical equation (5) to give rise to oscillations. (This is, admittedly, heuristic.) Its *irreducibility* is contingent on each state being persistent. In the long run:

$$Pr[X(t+1)=x] = \sum_{x'} Pr[X(t+1)=x \mid X(t)=x'] * Pr[X(t)=x'] \quad (8)$$

will vanish if and only if for every *persistent* $\mathbf{x}'$ the transition probability:

$$Q(x|x') = Pr[X(t+1) = x \mid X(t) = x'] \quad (9)$$

is zero. (Then there is no way *into* **x** from an irreducible persistent set.)

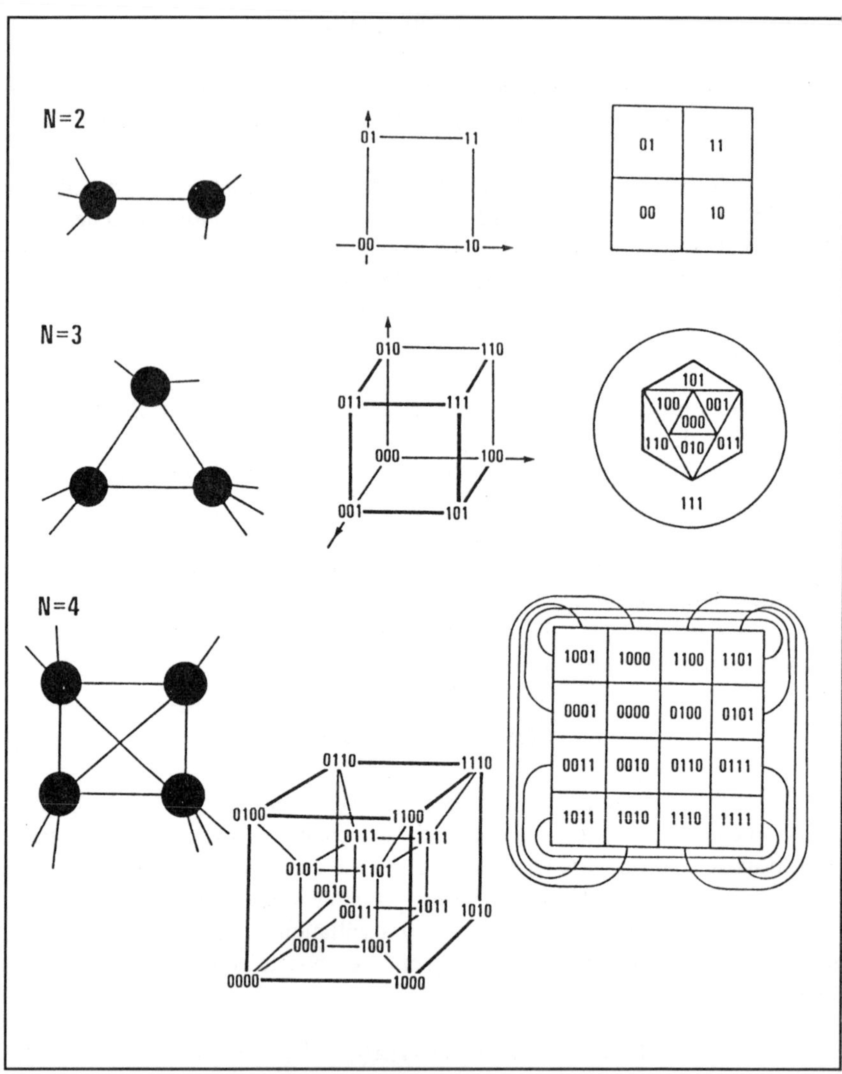

Figure 6.6 State transition maps for small systems of McCulloch-Pitts neurons.

The dynamical equation (5) restricts the steps of the random walk {**X**(t)} to *adjacent* vertices of the unit N-cube, depicted in fig. 6.6 for N = 2, 3, and 4. Let D(**x**,**x**′) be the Hamming distance as before (equation 4.7). Then D(**x**,**x**′) $\geq$ 2 implies Q(**x**|**x**′) = 0.

If $D(\mathbf{x},\mathbf{x}') = 1$, then $|x_j - x_j'| = \delta_{ij}$ for $i = K(t)$. The probability of a transition of distance one is:

$$Q(\mathbf{x}|\mathbf{x}') = Pr\left\{ K(t) = i \text{ and } \mathcal{H}\left[\sum_j W_{ij} x_j - \theta_i + z_i(t) \right] = x_i \right\} . \tag{10}$$

Since K(t) and $\mathbf{Z}(t)$ are independent, and K(t) is equiprobably any i, this is:

$$Q(\mathbf{x}|\mathbf{x}') = (1/N) Pr\left\{ \mathcal{H}\left[\sum_j W_{ij} x_j - \theta_i + Z_i(t) \right] = x_i \right\} \tag{11}$$

where the components of $\mathbf{x}$ are substituted for those of $\mathbf{x}'$ without loss of accuracy since these states differ only in the ith component and $W_{ii} = 0$. The probability on the right in equation (11) is readily evaluated for the case $x_i = 1$:

$$\begin{aligned} Pr[\sum_j W_{ij} x_j - \theta_i + Z_i(t) > 0] &= Pr\left[Z_i(t) \leq \sum_j W_{ij} x_j - \theta_i \right] \\ &= F\left(\sum_j W_{ij} x_j + Z_i \right) \end{aligned} \tag{12}$$

where $z_i = \mu_i - \theta_i$. Define:

$$q_i(\mathbf{x}) = F\left\{ \sum_{j=1}^{N} W_{ij} x_j + Z_i \right\} \tag{13}$$

and $q_i'(\mathbf{x}) = 1 - q_i(\mathbf{x})$. Then for every $\mathbf{x}$ and $\mathbf{x}'$ which differ only in the ith component we have:

$$Q(\mathbf{x}|\mathbf{x}') = \begin{cases} (1/N) q_i(\mathbf{x}) & \text{if } x_i = 1 \quad (x_i' = 0) \\ (1/N) q_i'(\mathbf{x}) & \text{if } x_i' = 1 \quad (x_i = 0) \end{cases} \tag{14a}$$

$$= (1/N)[x_i q_i(\mathbf{x}) + x_i' q_i'(\mathbf{x})]$$

(Recall $x_i, x_i' \in \{0,1\}$.) It remains to consider transitions of zero distance the probabilities of which fill out the diagonal of the $2^N \times 2^N$ stochastic matrix $\mathbf{Q}$. Evidently:

$$Q(x|x) = 1 - \sum_{x'} Q(x'|x) \quad ,$$

the summation ranging over all $\mathbf{x}'$ which differ from $\mathbf{x}$ by a single bit. It should only be necessary to refer to (14a) and reverse the roles of the state vectors to obtain:

$$Q(\mathbf{x}|\mathbf{x}) = 1 - (1/N)\sum_{i=1}^{N} [x_i' q_i(\mathbf{x}) + x_i q_i'(\mathbf{x})] \quad . \tag{14b}$$

Stationary Distribution

A distribution will be stationary if the number of items leaving a given range is equal to the number entering. Now return to equation (8) and assert the existence of the stationary distribution P(**x**) by writing:

$$P(\mathbf{x}) = \sum_{\mathbf{x}'} Q(\mathbf{x}|\mathbf{x}')P(\mathbf{x}') \quad . \tag{15}$$

Equation (15) follows from (8) when the distribution of **X**(t+1) is the same as that of **X**(t); and this is precisely the meaning of stationarity. Moreover, the existence of this stationary distribution guarantees the ergodicity that makes P(**x**) both the long run distribution of the state of the single system and the instantaneous distribution of a (canonical) ensemble of constitutionally identical systems. A more comprehensive treatment of the theory can be found in any good text on Markov process theory. Most such texts will be updates of the classic by Feller (1957). Once again, the only nonzero transition probabilities are for Hamming distance one and zero. Thus:

$$P(x) = \sum_{x' \neq x} Q(x|x')P(x') + Q(x|x)P(x)$$

and:

$$P(x)[1 - Q(x|x)] = \sum_{x' \neq x} Q(x|x')P(x') \quad .$$

Substituting equations (14) gives the more explicit form:

$$P(\boldsymbol{x})\sum_{i=1}^{N}[x_i' q_i(\boldsymbol{x}) + x_i q_i'(\boldsymbol{x})] = \sum_{i=1}^{N}(P(x_i';\boldsymbol{x})[x_i q_i(\boldsymbol{x}) + x_i' q_i'(\boldsymbol{x})]) \qquad (16)$$

with the substitution notation:

$$P(x_1';\mathbf{x}) = P(x_1', x_2, \ldots, x_N),$$

$$\ldots$$

$$P(x_i';\mathbf{x}) = P(x_1, \ldots, x_i', \ldots, x_N), \text{ etc.}$$

Problem 6.5: Consider a system of two McCulloch-Pitts neurons (fig. 6.7) with states $X_1 = X$ and $X_2 = Y$ and weights $W_{12} = a$ and $W_{21} = b$ not necessarily equal. The units are driven by independent white noise sources Z and Z′ having common distribution function $F(z) = 1 - F(-z)$.

Assuming the dynamics of equation (5), calculate the stationary distribution [the joint density of X and Y].

Solution:

(a) Begin with equation (15) for the stationary distribution $P(\mathbf{x})$. The network state vector is the random variable $\mathbf{X} = (X,Y)$ and its realizations are $\mathbf{x} = (x,y)$. Simplify the notation somewhat by omitting the comma and writing $P(xy)$ for $P(x,y)$. Thus $P(00)$ is the stationary probability that both units are OFF (and so on). The four components of:

$$\mathbf{P} = [P(00)\ \ P(01)\ \ P(10)\ \ P(11)]^T,$$

a column vector, are the unknowns in the system of linear equations:

$$\mathbf{P} = \mathbf{QP}\ ,$$

obtained from equation (15), in which $\mathbf{Q}$ is the transition matrix. The off-diagonal components $Q(\mathbf{x}|\mathbf{x}') = Q(xy|x'y')$ of $\mathbf{Q}$ are used to label the state transitions in fig. 6.8(a). Equation (11) for the transition probabilities now gives rise to four cases:

$$Q(1y|0y) = (\tfrac{1}{2})\Pr\{ay + Z > 0\},$$

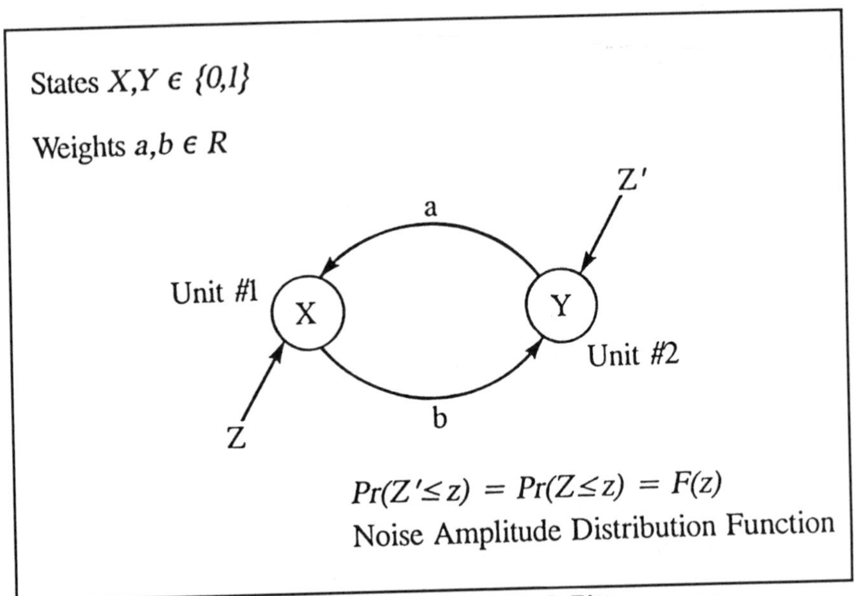

Figure 6.7 System of two McCulloch-Pitts neurons.

$$
\begin{aligned}
Q(0y \mid 1y) &= (\tfrac{1}{2})\Pr\{ay + Z \leq 0\}, \\
Q(x1 \mid x0) &= (\tfrac{1}{2})\Pr\{bx + Z' > 0\}, \\
\text{and } Q(x0 \mid x1) &= (\tfrac{1}{2})\Pr\{bx + Z' \leq 0\},
\end{aligned}
$$

from which the eight probabilities in fig. 6.8(b) are obtained. For examples (see page 126):

$$
\begin{aligned}
Q(11 \mid 01) &= (\tfrac{1}{2})\Pr\{a + Z > 0\} \\
&= (\tfrac{1}{2})\Pr(Z > -a) \\
&= (\tfrac{1}{2})[1 - \Pr(Z \leq -a)] \\
&= (\tfrac{1}{2})[1 - F(-a)] \\
&= F(a)/2
\end{aligned}
$$

and:

$$Q(00 \mid 01) = (\tfrac{1}{2})\Pr\{Z' \leq 0\} = F(0)/2 = \tfrac{1}{4}.$$

In the figure, the definitions:

$$A = F(a)$$

and:

$$B = F(b)$$

are invoked. In the sequel we use the additional notations:

$$\overline{A} = 1\text{-}A$$

and:

$$\overline{B} = 1\text{-}B.$$

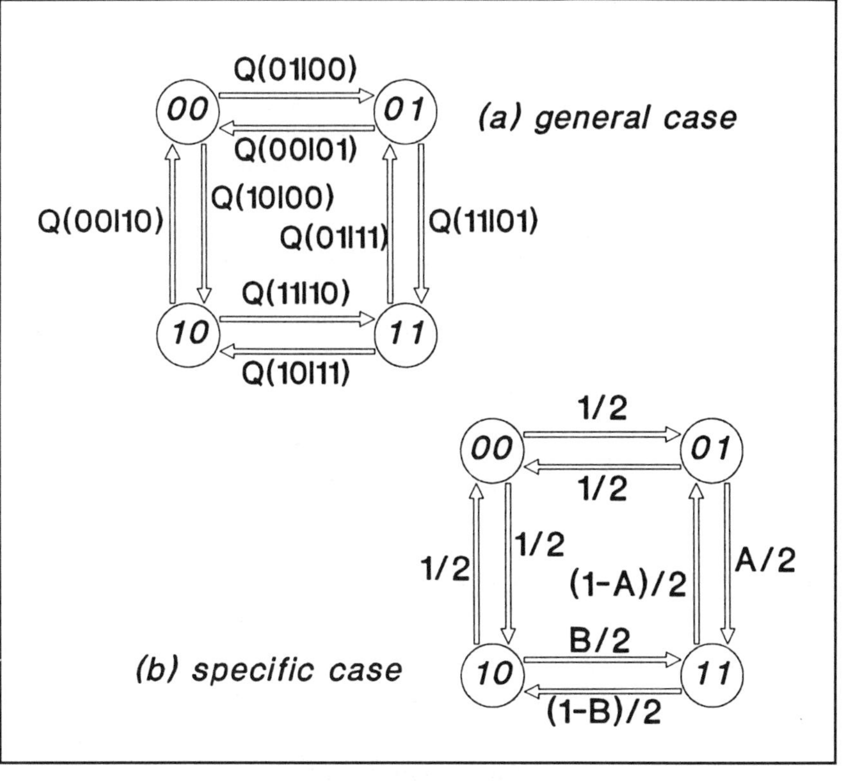

Figure 6.8 State transition diagrams showing the probabilities of transitions of Hamming distance one for the two-unit network (a) in general and (b) for the specifics of Problem 6.5.

The eight probabilities shown in fig. 6.8 are the nonzero, nondiagonal elements of **Q**. The diagonal elements, which concern "transitions of distance zero", are readily obtained from the state transition diagram:

$$Q(00|00) = 1 - Q(01|00) - Q(10|00) = \tfrac{1}{2}$$
$$Q(01|01) = 1 - (\tfrac{1}{2} + A)/2$$
$$Q(10|10) = 1 - (\tfrac{1}{2} + B)/2$$

$$Q(11|11) = 1 - (A + B)/2 .$$

Now the system **P** = **QP** (see equation 15) is expanded and rearranged to yield the following four simultaneous equations in the four unknown stationary probabilities:

$$\begin{aligned} P(00) &= \tfrac{1}{2} P(01) + \tfrac{1}{2} P(10) \\ P(01)*(A + \tfrac{1}{2}) &= \tfrac{1}{2}P(00) + P(11)A \\ P(10)\,(B + \tfrac{1}{2}) &= \tfrac{1}{2}P(00) + P(11)B \\ P(11)\,(A + B) &= P(01)A + P(10)B. \end{aligned}$$

Substituting for P(01) and P(10) in the right side of the first equation results in a direct proportionality between the first and last components, as:

$$P(11) = gP(00), \qquad \textbf{(17a)}$$

where the proportionality factor is:

$$g = \frac{(1-(\tfrac{1}{2})[1/(2A+1)+1/(2B+1)]}{A/(2A+1) + B/(2B+1)} \;. \qquad \textbf{(17b)}$$

The other two components can also be expressed as multiples of P(00):

$$P(01) = P(00)\frac{\tfrac{1}{2} + gA}{\tfrac{1}{2} + A} \qquad \textbf{(17c)}$$

$$P(10) = P(00)\frac{\tfrac{1}{2} + gB}{\tfrac{1}{2} + B}. \qquad \textbf{(17d)}$$

The stationary probabilities for any pair (a,b) of weights are easily found with the aid of a programmable calculator: Assume that P(00) is any positive number, find the sum of all four components, and then renormalize (by dividing all four components by their sum). As a quick "sanity check" of equations (17), verify that decoupling the neurons, as a = b = 0, results in all states being equiprobable: P(00) = P(01) = P(10) = P(11) = ¼.

If the foregoing algebra was not tedious enough, we can repeat the process starting from some assumptions which, though they may be heuristic, turn out to be entirely equivalent to those which led to equation (15). While it yields the same mathematical result, this alternative derivation leads to expressions which may be formally more transparent. Let the stationary probabilities form a 2-by-2 contingency table:

$$\boldsymbol{P} = \begin{pmatrix} p_{00} & p_{01} \\ p_{10} & p_{11} \end{pmatrix} ,$$

where p_{xy} is the same as P(x,y). Such tables, in general, give rise to four marginal probabilities that are obtained as row sums and column sums. In the present case, the marginals reflect the long run statistics of the individual units considered by themselves:

$$\begin{aligned} \Pr(X = 0) &= q_0 = p_{00} + p_{01} \\ \Pr(X = 1) &= q_1 = p_{10} + p_{11} \\ \Pr(Y = 0) &= r_0 = p_{00} + p_{10} \\ \Pr(Y = 1) &= r_1 = p_{01} + p_{11} \end{aligned}$$

The very definition of conditional probability gives us:

$$\Pr(X=x,\ Y=y) = \Pr(X=x \mid Y=y)\Pr(Y=y)$$

and

$$\Pr(X=x,\ Y=y) = \Pr(Y=y \mid X=x)\Pr(X=x) .$$

Since the common left side of the last two statements was denoted p_{xy}, they combine to give:

$$p_{xy} = (½)\Pr(X=x \mid Y=y)r_y + (½)\Pr(Y=y \mid X=x)q_x \qquad \textbf{(18)}$$

after substituting the marginals in the appropriate places. Let the conditional probability of having X=1, conditioned on Y=y, be regarded as the likelihood that the first unit is turned ON given that we inspect its state at some time when the second unit is in state y. This likelihood is:

$$\Pr(X=1 \mid Y=y) = \Pr(Z+ay > 0) = \Pr(Z > -ay) = 1 - F(-ay) .$$

The conditional probability of having X = 0, then, is obviously:

$$\Pr(X=0 \mid Y=y) = 1 - F(ay) .$$

By the formal symmetry of the problem:

$$\Pr(Y=1 \mid X=x) = F(bx)$$

and:

$$\Pr(Y=0|X=x) = 1 - F(bx) .$$

With the same definitions of A and B, $\bar{A}$ and $\bar{B}$, as noted above, use the last four statements to enumerate the eight cases of the conditional probability:

$$\begin{aligned}
\Pr(X=0|Y=0) &= F(0) = ½ \\
\Pr(X=0|Y=1) &= 1\text{-}F(a) = \bar{A} \\
\Pr(X=1|Y=0) &= F(0) = ½ \\
\Pr(X=1|Y=1) &= F(a) = A \\
\Pr(Y=0|X=0) &= F(0) = ½ \\
\Pr(Y=0|X=1) &= 1\text{-}F(b) = \bar{B} \\
\Pr(Y=1|X=0) &= F(0) = ½ \\
\Pr(Y=1|X=1) &= F(b) = B.
\end{aligned}$$

Substitute these conditional probabilities into the four cases of equation (18) to obtain four linear equations in the unknowns p_{xy}, q_x, and r_y:

$$\begin{aligned}
p_{00} &= r_0/4 + q_0/4 \\
p_{01} &= r_1\bar{A}/2 + q_0/4 \\
p_{10} &= r_0/4 + q_1\bar{B}/2 \\
p_{11} &= r_1 A/2 + q_1 B/2
\end{aligned}
\quad
\begin{cases}
A = F(a) \\
\bar{A} = 1 - A \\
B = F(b) \\
\bar{B} = 1 - B.
\end{cases}
\tag{19}$$

Let us pause here to check that we have indeed a probability density:

$$\begin{aligned}
\sum_{x=0}^{1}\sum_{y=0}^{1} p_{xy} &= ½\, r_0 + ½\, q_0 + ½\, q_1(B+\bar{B}) + ½\, r_1(A+\bar{A}) \\
&= ½\,(r_0+q_0) + ½\,(r_1+q_1) \\
&= ½\,(r_0+r_1) + ½\,(q_0+q_1) \\
&= 1.
\end{aligned}$$

Next we calculate the marginals:

$$\begin{aligned}
r_0 &= p_{00} + p_{10} \\
&= ¼(r_0+q_0) + (¼r_0 + ½q_1\bar{B}) \\
&= 2(¼q_0+½q_1\bar{B}) \\
r_0 &= ½ - q_1(B - ½).
\end{aligned}$$

And, by symmetry:

$$q_0 = p_{00} + p_{01} = ½ - r_1(A - ½).$$

Now substitute 1 - r_0 for r_1 in the equation for q_0:

$$q_0 = 1/2 - [1/2 - q_1(B-1/2](A-1/2)$$

$$= 1/2 - [1/2 + (1-q_0)(A-1/2)(B-1/2)]$$

$$= \frac{1/2 - (A-1/2)B}{1 - (A-1/2)(B-1/2)}.$$

Again, by symmetry:

$$r_0 = \frac{½ = (B-½)A}{1 - (A-½)(B-½)}.$$

Finally, we calculate the joint density: Define D = 1 - (A-½)(B-½) and write the above results as:

$$q_0 = (½-(A-½)B)/D \quad and \quad r_0 = (½-(B-½)A)/D \ . \qquad \textbf{(20a)}$$

Observe that:

$$p_{00} = ¼(r_0+q_0) = [1-(A - ½)B - (B - ½)A]/4D.$$

Also:

$$q_1 = (A + ½)/2D \qquad and \qquad r_1 = (B + ½)/2D$$

(20b)

Recalling the first of equations (19) for p_{10}, and substituting for r_0 and q_1 in accordance with (20), find:

$$p_{10} = \frac{1-2(B - ½)A + ½(A - B)}{4D} \quad .$$

Similarly:

$$P_{01} = \frac{1 - 2(A - ½)B + ½(B - A)}{4D}$$

Finally:

$$p_{11} = 1 - p_{00} - p_{10} - p_{01}$$

$$= 1 - \frac{3}{4D}(1 - (A - ½)B - (B - ½)A)$$

In summary, the joint density P(x,y) is given by:

$$\begin{bmatrix} p_{00} & p_{01} \\ p_{10} & p_{11} \end{bmatrix} = \frac{1}{4D} *$$

$$\begin{bmatrix} 1 - (A-½)B - (B-½)A & 1 - 2(A-½)B + ½(B-A) \\ 1 - 2(B-½)A + ½(A-B) & 4D - 3(1 - (A-½)B - (B-½)A \end{bmatrix}$$

(21)

where D = 1 - (A - 1/2)(B - 1/2), A = F(a) and B = F(b). No conditions of symmetry have been placed on the weight matrix and this result is quite general.

A numerical example is: a = 1, b = -1, Z ~ LOGIS(0,T) and 1/T = ln(3). With these parameters:

A = F(a) = 1/(1 + exp(-ln3)) = 3/4
B = F(b) = 1/(1 + exp(+ln3)) = 1/4 .

and the predicted joint density is:

$$\begin{bmatrix} p_{00} & p_{01} \\ p_{10} & p_{11} \end{bmatrix} = \begin{bmatrix} 9/34 & 5/34 \\ 13/34 & 7/34 \end{bmatrix} = \begin{bmatrix} .265 & .147 \\ .382 & .206 \end{bmatrix} .$$

Since it is not terribly obvious from looking at the formal expressions, the student may want to show, by writing a simple program, that equations (21) give precisely the same numbers as (the normalized) equations (17).

If a and b are equal, the weight matrix is symmetric and B = A = F(a) and:

$$\begin{bmatrix} p_{00} & p_{01} \\ p_{10} & p_{11} \end{bmatrix} = \frac{1}{D} \begin{bmatrix} 1+A-2A^2 & 1+A-2A^2 \\ 1+A-2A^2 & 2A^2+A \end{bmatrix} = \frac{1}{D'} \begin{bmatrix} \bar{A} & \bar{A} \\ \bar{A} & A \end{bmatrix}$$

where $D = 4(1-(A-1/2)^2)$, $D' = 3-2A$ and $\bar{A} = 1 - A$. Since $p_{11} / p_{00} = p_{11} / p_{01} = p_{11} / p_{10} = A/A$, the joint density can be rewritten as:

$$\frac{\bar{A}}{D'} \begin{bmatrix} 1 & 1 \\ 1 & A/\bar{A} \end{bmatrix} = \frac{1}{3+e^{\log(A/\bar{A})}} \begin{bmatrix} e^0 & e^0 \\ e^0 & e^{\log(A/\bar{A})} \end{bmatrix}$$

dispensing with the matrix notation, the joint density can be written as:

$$P(x,y) = \frac{\exp(xy \; log \; \frac{A}{1-A})}{\sum_{x,y} \exp(xy \; log \; \frac{A}{1-A})} ,$$

$x,y \in \{0,1\}$.

In the special case of logistic noise, the substitution $A = F(a) = 1/(1 + \exp(-a/T))$ leads to:

$$P(x,y) = \frac{e^{xya/T}}{\sum_{x,y} e^{xya/T}} ,$$

which is the Boltzmann distribution.

For this system of two neurons, the symmetry of the weight matrix (b=a) gives rise to a stationary distribution with:

$$P_{xy} \propto \exp(-xy \log\left[\frac{1}{F(a)} - 1\right]) .$$

The logistic distribution function with scale parameter T has the unique property that:

$$1/F(a) - 1 = \exp(-a/T).$$

Thus:

$$P_{xy} \propto \exp([xya] / T)$$

when the noise is logistic. This is a Boltzmann distribution since, in the present problem, $H(x,y) = -xya$ is the energy of state (x,y).

Problem 6.6: A pair of McCulloch-Pitts neurons with states s_1 and s_2 are linked by weight matrix **T**: $T_{ij} = a$ $(i \neq j)$ and $T_{ii} = 0$. The units are driven by independent white noise sources with common distribution function F. If $F(a) = 0.75$, what is the stationary distribution $P(s_1,s_2)$?

Solution: Figure 6.9 sketches the computation of the transition probabilities and their use in finding directly a solution to equation (15). The result is the same as would be obtained by application of equation (18).

Problem 6.7: Consider a network of three McCulloch-Pitts neurons with reciprocal links of strength +A and -A as shown. Calculate the stationary probabilities of the eight states, assuming a Boltzmann distribution at temperature $T = A/\ln(2)$. Verify this by programming the dynamical equation (5) and observing the relative frequencies with which the states are visited. Does synchronous dynamics also produce a Boltzmann machine?

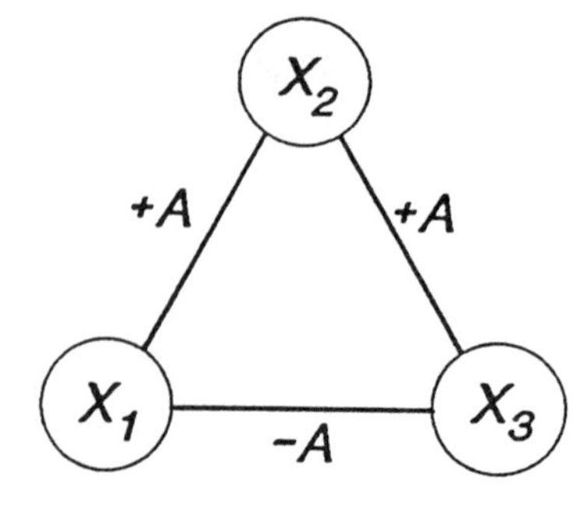

Solution: The partition function is a sum over the eight states, which share three energy levels:

$$Z(T) = 4 + 3e^{A/T} + e^{-A/T} = 10.5 .$$

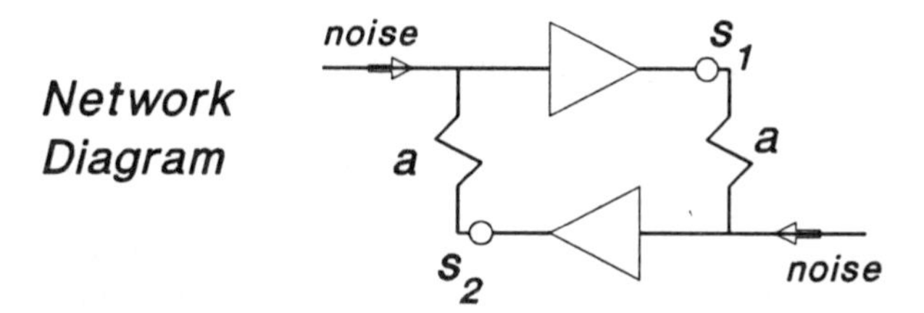

Transition probabilities:

$$Q(s + ds \mid s) = \begin{cases} (1/N)F(s'Tds) & \text{if } ds = +\delta_{ij} \\ (1/N)[1 - F(s'Tds)] & \text{if } ds = -\delta_{ij} \end{cases}$$

$F(a) = 3/4 \Longrightarrow$

$$Q(01|00) = Q(10|00) = F(0)/2 = 1/4$$
$$Q(11|01) = Q(11|10) = F(a)/2 = 3/8$$
$$Q(00|01) = Q(00|10) = [1 - F(0)]/2 = 1/4$$
$$Q(01|11) = Q(10|11) = [1 - F(a)]/2 = 1/8$$

Stationary distribution:

$$\begin{bmatrix} 1/2 & 1/4 & 1/4 & 0 \\ 1/4 & 3/8 & 0 & 1/8 \\ 1/4 & 0 & 3/8 & 1/8 \\ 0 & 3/8 & 3/8 & 3/4 \end{bmatrix} \begin{bmatrix} 1/6 \\ 1/6 \\ 1/6 \\ 1/2 \end{bmatrix} = \begin{bmatrix} 1/6 \\ 1/6 \\ 1/6 \\ 1/2 \end{bmatrix}$$

Generalized Boltzmann distribution:

$$\frac{\begin{bmatrix} 1 - F(a) & 1 - F(a) \\ 1 - F(a) & F(a) \end{bmatrix}}{3 - 2F(a)} = \begin{bmatrix} 1/6 & 1/6 \\ 1/6 & 1/2 \end{bmatrix}$$

Figure 6.9 Computation of transition probabilities.

Where T has been set equal to A/ln(2). Thus the state of highest energy h
probability:

$$P(1,0,1) = e^{-A/T}/Z = .5/10.5 = 0.048;$$

and the states of least energy have probability:

$$P(1,1,0) = e^{A/T}/Z = 2/10.5 = 0.19 .$$

The BASIC program (Appendix E) should not be hard to interpret. Note th
line 160 defines a (roughly) GEOmetric random variable with a mean of abo
10 ln(2) $\approx$ 7; and the network state is sampled every GEO time steps. Th
random sampling will keep the statistics from being slanted by periodicity whic
cannot arise with the stochastic dynamics of equation (5). When th
synchronous dynamics is used the random sampling serves a worthwhil
purpose.

A different BASIC program was written by Su[1] to provide synchronou
and asynchronous options. In the three McCulloch-Pitts neuron network wi
Glauber dynamics there are eight possible states. Su calculated the probabilit
distribution through a simulation of 10,000 random samples. His results a
summarized below in figures 6.10 and 6.11.

While the observed distribution with sample size ten thousand is almo
indistinguishable from the theoretical in the first case (fig. 6.10), th
synchronous dynamics leads to a decidedly different set of statistic
(figure 6.11). We shall reflect upon this disparity again in the last sectio
which concerns the Gibbs-Markov equivalence.

Glauber's "Master Equation"

A more intuitive derivation of equation (16) for the stationar
distribution was offered by Glauber (1963) in his study of the time-depende
statistics of the Ising model. On the left in (16) we have the probability of stat
x multiplied by the sum of all the *exit* probabilities $Q(\mathbf{x}'|\mathbf{x})$. Multiply both side
by the number M of constitutionally identical but physically independent system
(networks). Then $MP(\mathbf{x})$ is the mean number of systems in state **x**. Als
multiply both sides by a suitable rate parameter Υ:

---

[1]At the time when this work was done, Mark Su was a junior at Blair Hig
School, Montgomery County, Maryland.

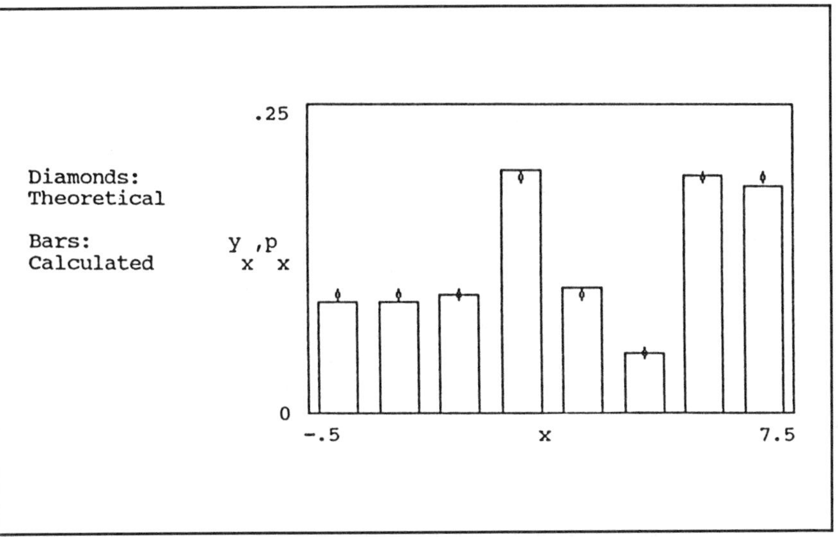

Figure 6.10 Calculated stationary distribution and theoretical Boltzmann distribution - Glauber dynamics.

$[x_i'q_i(\mathbf{x}) + x_iq_i'(\mathbf{x})]\Upsilon$ = the rate of conversion of state $\mathbf{x}$ into state $\mathbf{x}' = (x_1 \ldots x_i' \ldots x_N)$.

Similarly:

$[x_iq_i(\mathbf{x}) + x_i'q_i'(\mathbf{x})]\Upsilon$ = the rate of conversion of state $\mathbf{x}'$ into state $\mathbf{x}$.

Now the product $\Upsilon MP(x_i';\mathbf{x})Q(\mathbf{x}|\mathbf{x}')$ is the number of systems per unit time converted into $\mathbf{x}$ from $\mathbf{x}'$ and similarly for the reverse process. Then the right side of equation (16) is the flux into state $\mathbf{x}$ and the left side is the outward flux. Clearly the time-dependent probability $p(\mathbf{x},t)$ satisfies the differential equation:

$$\Upsilon M[dp(\mathbf{x},t)/dt] = [\text{flux into } \mathbf{x}] - [\text{flux out of } \mathbf{x}].$$

With the simplified notations $q_i(x_i)$ and $q_i(x_i')$ for the conversion rates out and in, respectively, i.e., with:

$$q_i(x_i) \Leftrightarrow \Upsilon x_i'q_i(\mathbf{x}) + \Upsilon x_iq_i'(\mathbf{x})$$

and:

$$q_i(x_i') \Leftrightarrow \Upsilon x_iq_i(\mathbf{x}) + \Upsilon x_i'q_i'(\mathbf{x}),$$

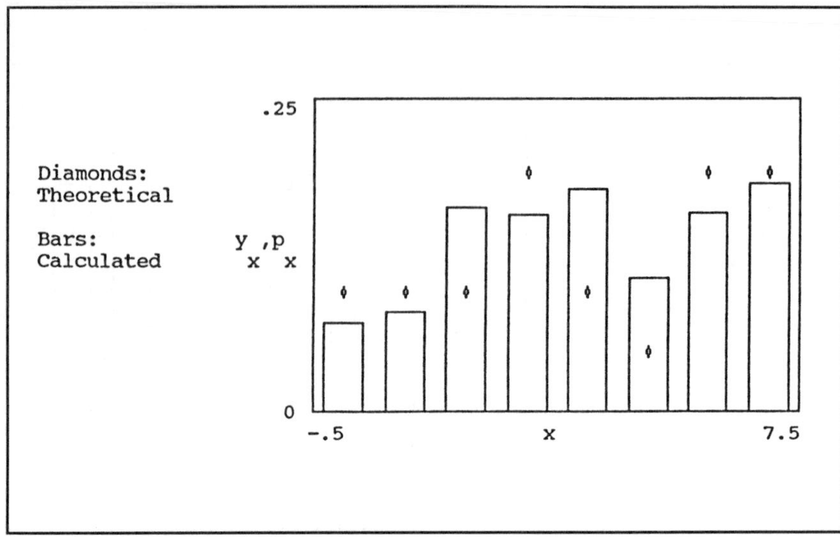

Figure 6.11 Calculated stationary distribution and theoretical Boltzmann distribution - synchronous dynamics.

the differential equation can be rewritten as:

$$dp(\mathbf{x},t)/dt = \sum_{i=1}^{N} p(x_i';\mathbf{x},t)q_i(x_i') - p(\mathbf{x},t)\sum_{i=1}^{N} q_i(x_i) \quad , \tag{24}$$

where $p(x_i';\mathbf{x},t)$ is the time-dependent probability of the state$(x_1 \ldots x_i' \ldots x_N)$. This (24) is Glauber's master equation in essentially the same notation as the original. Figure 6.12 starts from the master equation for $p(\mathbf{s},t)$, assuming sign neurons (Ising spins) so that $s_i'=-s_i$, and goes back to equation (16) by setting the derivative to zero. The result is 2^N linear equations in as many unknowns, the stationary probabilities $\{P(\mathbf{x})\}$.

Problem 6.8 (Detailed Balance): With reference to fig. 6.12, suppose that $P(s_1,\ldots,s_N)q_i(s_i) = P(s_i,\ldots,-s_i,\ldots,s_N)q_i(-s_i)$ for each i and s. What does this imply regarding the *odds ratio*:

$r_i(\mathbf{s}) = P(s_1 \ldots s_i = +1 \ldots s_N)/P(s_1 \ldots s_i = -1 \ldots s_N)$?

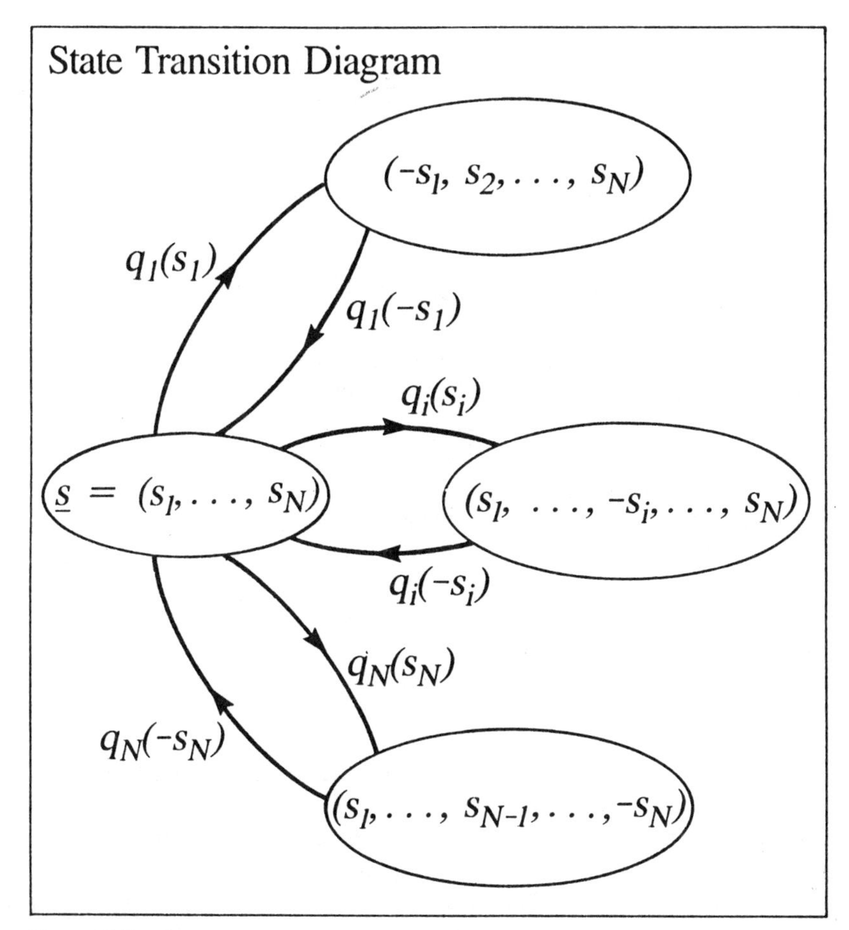

Equilibrium:

$$[\sum_i q_i(s_i)]p(\underline{s}) = \sum_i [q_i(-s_i)p(s_1, \ldots, -s_i, \ldots, s_N)]$$

Figure 6.12 The Master Equation of R. J. Glauber.

Assume no thresholds or biases.

Solution: In general:

$$r_i(s) = \frac{F(\sum_j W_{ij}s_j)}{[1 - F(\sum_j W_{ij}s_j)]} \, .$$

In the specific case of logistic noise with zero mean and scale parameter T this is $\ln[r_i(s)] = \Sigma_j W_{ij}s_j / T$.

The Gibbs-Markov Equivalence

Dobrushin (1968) generalized the concept of the Markov chain to higher-dimensional lattices. (Recall the first part of Chapter 4). As a result of this mathematical exercise, captured to some extent in fig.s 6.13 and 6.14, the Markov chain becomes the simplest Markov field. In a Markov field, the probability that a lattice site assumes a particular state is directly influenced only by the neighboring sites. The sites that are not neighbors can only influence the site in question indirectly. In the Markov chain, the nearest neighbors of the site ("present time") are two in number ("before" and "after"). Dobrushin's work was treated with some dismay in the West (as were others of his difficult papers); and the eventual outcome was the elaboration of the Gibbs-Markov equivalence by Spitzer (1971) and Moussouris (1974).

Following Dobrushin, Spitzer considered Euclidean lattices without restriction on dimensionality or on the set of discrete states that the single site can assume. This set **S** of possible states is the domain of X. When **S** contains only the two elements -1 and +1, the Markov field is just a simple Ising model in an arbitrary number of dimensions. Enlarging **S** to include any countable repertoire of values, let x_i be the value of the site located at i and so on. Define a pair potential $\phi(i,j)$ that has three properties:

$$\phi(i,j) = \phi(j,i) \quad \text{(symmetry)},$$

$$\phi(i,j) = 0 \text{ if } |j-i| > 1 \quad \text{(the nearest neighbor property)},$$

and

$$\phi(i,j) = \phi(0,j-i) \quad \text{(homogeneity)}.$$

The pair potentials are the elements of the interaction matrix **J** of a generalized Ising model. They can also be the weights in a matrix W describing connectivity in a Euclidean lattice of abstract neurons that can assume different levels of activation.

Markov Chain in which the probability that a node assumes given value is decided by the nodes immediately before and after.

$$\cdots \quad i-1 \quad i \quad i+1 \quad i+2 \quad \cdots$$

$$Pr(X_i = x \mid X_0, \ldots, X_{i-1}, X_{i+1}, \ldots) = Pr(X_i = x \mid X_{i-1}, X_{i+1})$$

2D Markov Field in which the probability that a node assumes given value is decided by its four nearest neighbors.

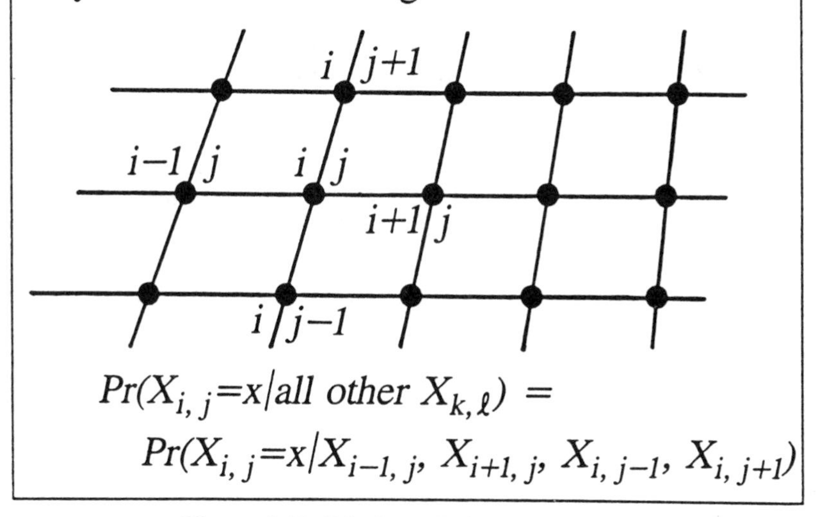

$$Pr(X_{i,j} = x \mid \text{all other } X_{k,\ell}) = Pr(X_{i,j} = x \mid X_{i-1,j}, X_{i+1,j}, X_{i,j-1}, X_{i,j+1})$$

Figure 6.13 Markov chains and fields.

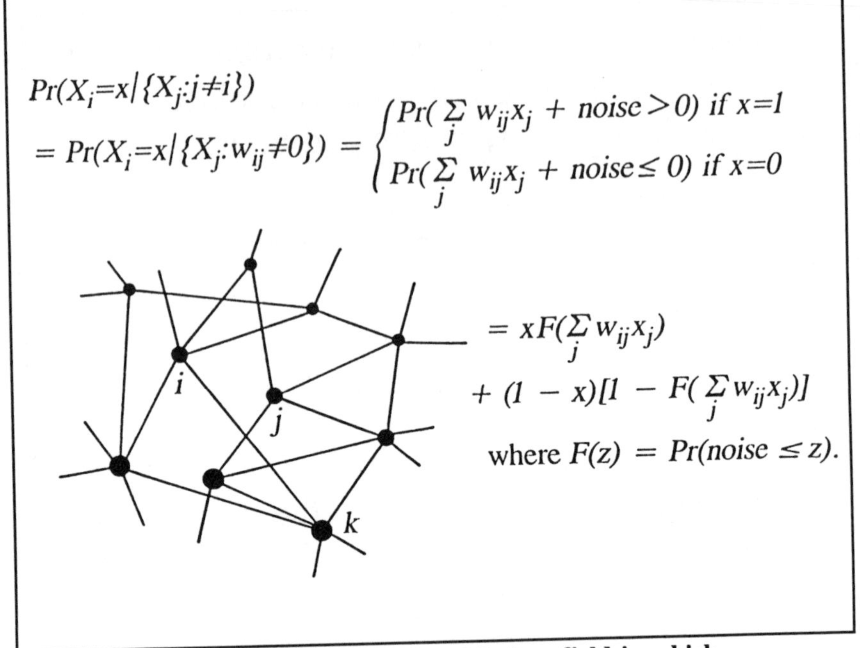

$Pr(X_i = x | \{X_j : j \neq i\})$
$= Pr(X_i = x | \{X_j : w_{ij} \neq 0\}) = \begin{cases} Pr(\sum_j w_{ij}x_j + noise > 0) \text{ if } x = 1 \\ Pr(\sum_j w_{ij}x_j + noise \leq 0) \text{ if } x = 0 \end{cases}$
i
j
k
$= xF(\sum_j w_{ij}x_j)$
$+ (1 - x)[1 - F(\sum_j w_{ij}x_j)]$
where $F(z) = Pr(noise \leq z)$.

Figure 6.14 N-dimensional Markov field in which proximity is defined by connectivity.

Let $\mathbf{x} = (x_i, x_j, x_k, \ldots)$ denote a configuration obtained by listing the values assumed by sites at the locations i, j, k, and so on. The random field (obtained by assigning probabilities to all the states) is Gibbsian if:

$$Pr(X = x) = \frac{e^{-\Phi}}{Z}$$

where:

$$\Phi(x) = -\sum_{i<j} \phi(i,j)x_i x_j$$

and Z is the normalizing divisor. (Note: The sum that yields the gross potential Φ is over all pairs of sites as before.) Spitzer proved two theorems: (i) Every Gibbs random field is a Markov random field and (ii) the converse.

Moussouris took the next step by shedding the notion of a Euclidean lattice. Fig. 6.15 shows nine sites that are the vertices of an undirected graph whose edges represent mutual interactions. To keep from having to compile a book of synonyms, let the sites and interactions define a network so that A, B, and C can be called subnets and their vector values ($\mathbf{X}_A$, $\mathbf{X}_B$, and $\mathbf{X}_C$) are the substates comprising an overall state vector $\mathbf{X} = (\mathbf{X}_A, \mathbf{X}_B, \mathbf{X}_C)$. Note that A and C are disjoint subnets having no direct links between them. Subnet B is broken down into two parts: BA communicates only with subnet A and BC only with subnet C.

Let **x** be any realization of the network state and use the operator P to take its probability:

$$\Pr(\mathbf{X} = \mathbf{x}) = P(\mathbf{x}) = P(\mathbf{x}_A\ \mathbf{x}_B\ \mathbf{x}_C) \text{ etc.}$$

That subnets A and C do not interact directly means conditioning the distribution of A on B and C is the same as conditioning on B alone:

$$P(x_A|x_B x_C) = P(x_A|x_B) \quad . \tag{25}$$

This is the Markov assumption, although other ways of stating it might be more illuminating. By using (first) the definition of conditional probability and (second) the Markov property (23), one obtains:

$$\frac{p(x_A x_B x_C)}{p(x_B x_C)} = \frac{p(x_A x_B)}{p(x_B)}. \tag{26}$$

Multiply both sides by $P(\mathbf{x}_A\mathbf{x}_B)$ and note that:

$$P(x_C|x_A x_B)P(x_A x_B) = P(x_A x_B x_C).$$

Then (25) becomes:

$$P(x_A x_B)P(x_B x_C) = P(x_A x_B x_C)\,P(x_B) \quad .$$

One can write this using dots (.) to indicate summations over all possible substates:

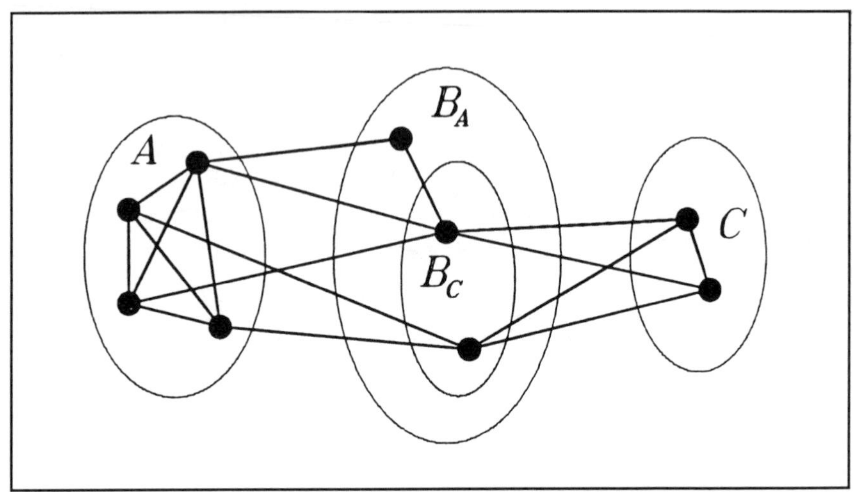

Figure 6.15 A network of nine sites which form three subnets (A, B, and C). Subnet B is subdivided into B_A and B_C, the "environments" of A and C, respectively.

$$P(x_A x_B .)P(. x_B x_C) = P(x_A x_B x_C)P(. x_B .) \quad .$$

Moreover, as Moussouris showed, we can replace the dot in the A-position with any substate $\mathbf{x'}_A$ that is different from $\mathbf{x}_A$, etc. This procedure yields the identity:

$$P(x_A x_B x_C)P(x'_A x_B x'_C) = P(x_A x_B x'_C)P(x'_A x_B x_C) \quad . \tag{27}$$

In this context it would be natural to define a function $\mathbf{\Phi} = -\ln(P)$, even without the prompting of statistical physics, simply as a device for converting the multiplicative Markov condition (27) into a linear form:

$$\Phi(x_A \; x_B \; x_C) + \Phi(x'_A \; x_B \; x'_c) \tag{28}$$

$$= \Phi(x_A \; x_B \; x'_C) + \Phi(x'_A \; x_B \; x_C) \quad .$$

Example: Let the network have just three sites (A=1, B=2 and C=3) each of which is occupied by one McCulloch-Pitts neuron. Let the pairwise interactions be described by these weights:

$$W = \begin{bmatrix} 0 & a & 0 \\ a & 0 & c \\ 0 & c & 0 \end{bmatrix}$$

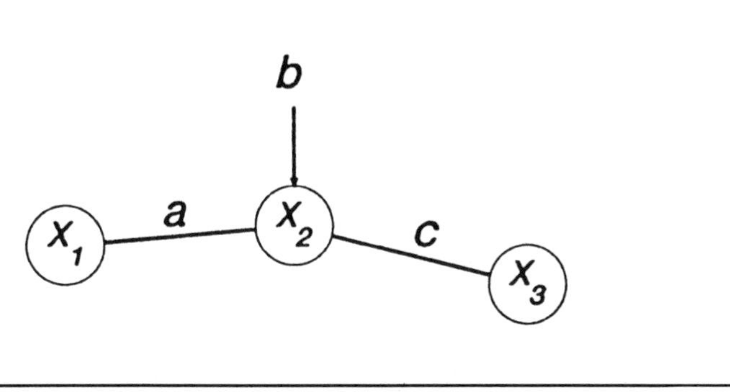

Figure 6.16 The subnets of the previous figure are reduced to single McCulloch-Pitts neurons.

ject a constant bias of +b into unit #2. It is easy to verify that (28) is tisfied when the computational energy H of the network state is substituted for e potential function. Let $(\mathbf{x}_A \mathbf{x}_B \mathbf{x}_C) = (1,1,1)$ for the sake of argument. Then 8) becomes:

$$\Phi(1,1,1) + \Phi(0,1,0) = \Phi(1,1,0) + \Phi(0,1,1) \quad .$$

ne network diagram in fig. 6.16 shows that the states in question have these ergies:

H(1,1,1) = -a-b-c	H(1,1,0) = -a-b
H(0,1,0) = -b	H(0,1,1) = -c-b

etting Φ = H, the above equation is satisfied.

Moussouris was not satisfied with just transplanting the Markov field om Euclidean space into the abstract realm of graph theory. He also perseded pair potentials with "clan potentials". The clans are what remain

after reducing the network to connected subnets and reducing every netwo
state to a corresponding collection of substates that contain no inacti
(zero-valued) sites. The clans can overlap. The simplest instance of the cl
is (obviously) the pair. A sum over all active clans in a system that has on
pairwise interaction is a weighted sum of products X_iX_j. Such a produ
vanishes when either site (i or j) is inactive. In addition, the pair contribut
nothing to the total potential unless i and j are connected with nonzero weigh
Similarly, in a system where the clans are triples, the active clans would exi
only when three connected sites are all active. In this way Markov fields ca
probably subsume a range of "higher order networks".

The Gibbs-Markov equivalence assures us that the states of a symmetr
neural network occur with relative frequencies that vary exponentially with the
potentials -- which are found by summing up the contributions made by all th
active clans. It rests on the assumption that the activation of a given unit com
from a stochastic mapping in which the activations of connected units and th
strengths of those connections are the controlling parameters. Accordingly,
should not matter what the details of the mapping are. (Such details include th
shape of the noise amplitude probability distribution that generates th
Gibbs/Boltzmann distribution of the states through the Glauber dynamic
equation.) The advanced student will want to take the last step in provin
theorem 2 by making the substitution:

$$P(\boldsymbol{x}) = \exp \sum_{i<j} \phi_{ij}\ (x_i x_j)$$

in equation (16), "cranking out" the pair potentials ϕ_{ij} as functions of th
network parameters. An additional challenge can be accepted by solving th
problem for a generic S-shaped noise amplitude probability distribution functio
F(z)--instead of assuming logistic noise as prescribed in the theorem. Appendi
B, which illustrates the Gibbs/Boltzmann distribution of the states of a four-un
network, offers a solution to the general case that should be treated with extrem
caution.

Additional Problems

6.9 A system of two sign neurons (Ising spins) has weight matri

$\boldsymbol{W} = \begin{pmatrix} 0 & a \\ b & 0 \end{pmatrix}$, $a > 0$, with no thresholds. If independent white noise source

drive the neurons, Glauber's master equation leads to the state transitio
diagram:

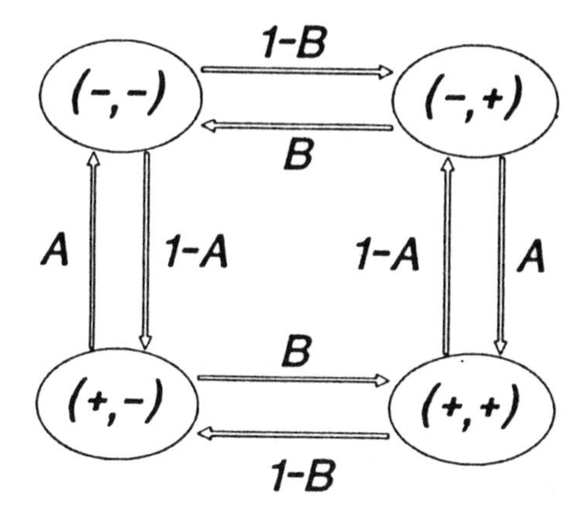

where A=F(a), B=F(b), and F(z) = Pr(noise $\leq$ z) = 1 - F(-z). What is the stationary (equilibrium) distribution of the four states? Specialize the result to the case a + b =0. Then let a=b and make the assumption of logistic noise, F(z) = 1/[1-exp(-z/T)].

6.10 True or false: The Boltzmann machine uses synchronous dynamics with a symmetric weight matrix and a sigmoidal response function (paraphrasing Grossberg 1988).

6.11 Consider the Glauber dynamical equation of a spin system with noise:

$$S_i(t+1) = \begin{cases} sign\left[\sum_j J_{ij} \; S_j(t) + \theta_i + \Xi(t) \right] & if \;\; i=K(t) \\ S_i(t) & otherwise \end{cases}$$

where **J** is a real, symmetric NxN matrix with $J_{ii}=0$, {K(t)} is a sequence of random integers uniformly distributed on the first N, and {ξ(t)} is a sequence of i.i.d. random variables with distribution function F(ξ) = Pr[Ξ(t)$\leq$ ξ] for every t. Assume F(-z) = 1 - F(z).

(a) Suppressing the time dependence after setting i=K(t), rewrite the dynamical equation in the form:

$$S_i = \begin{cases} +1 \ \textit{with probability } p \\ -1 \ \textit{with probability } 1-p \end{cases}$$

where p is a function of the *energy gap*:

$$\Delta H_i = H(\ s_1,\ldots,s_i = +1,\ldots,s_N\ ) - H(\ s_1\ldots s_i = -1\ldots s_N\ )$$

(b) What is $p(\Delta H_i)$ when $F(z) = 1/[1+\exp(-z/T)]$?

(c) For a particular i, let $J_{ij}=0$ for every j. What is the averag value of S_i? Assume the same form of F as in part (b).

6.12 A symmetric net of N+1 McCulloch-Pitts neurons is depicte here, with a "cloud" of N units fully interconnected by weights of strength -(≤ 0, and one unit, the $(N+1)^{st}$, which is linked to all the others by positiv weights of strength $A \geq 0$.

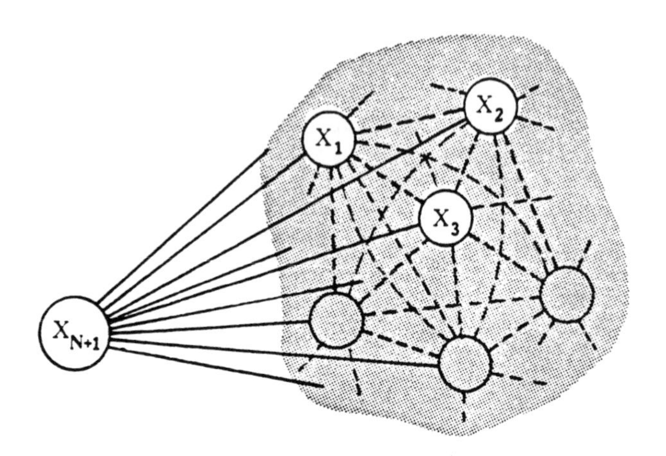

(a) Write an exact expression for the partition function, assuming C=0. Not that there are 2^N states with $X_{N+1} = 1$ and the same number of states wit the $N+1^{st}$ unit OFF. Each state is characterized by having k active and N- inactive (OFF) units. Recall:

$$\sum_{k=0}^{N} \binom{N}{k} = 2^N \quad .$$

(b) Now assume that $\beta C = 1$, where β is reciprocal temperature, and $A/C = a > 0$. Recalculate the partition function.

(c) With parameters fixed as in (b), the probability of a given state at thermal equilibrium clearly depends only on the number k of active units. Sketch p(k;a), the probability of k active units, for two or three values of a. (If no computational aids are available, merely try to capture the trend followed by the distribution as a increases.)

6.13 Recall Problem 3.12 and add to the bias of each unit a random part with amplitude distribution function $F(z) = 1-F(-z)$. In the limit as $A/C \rightarrow 0$, assume that only $N+1 = 6$ states have non-negligible probability. Sketch the state transition diagram and deduce the stationary distribution assuming the "detailed balance" of problem 6.8.

7. Escaping Local Minima To Improve Solution Quality

This chapter is to explore the third fundamental principle of collective computation (fig. 7.1), a principle of "computational neurodynamics (ND)." We saw in chapter 3 that the state of the Hopfield net converges (*in probability*) to one of the local minima of the computational energy. This claim was contingent upon the assumption of asynchronous dynamics with time-invariant weights, thresholds, and biases. In the preceding chapter, a randomly-varying component was added to the input of each binary unit. The addition of this "synaptic noise" was functionally equivalent to the use of stochastic neurons that respond to their fixed inputs by turning ON or OFF with probabilities that are functions of an energy gap. Again assuming asynchronous (Glauber) dynamics, the network takes a random walk through its state space, converging (*in distribution*) to the Gibbs/Boltzmann law, which describes systems in thermal equilibrium at constant temperature. In case chapter 6 did not belabor the point sufficiently, the Second Principle of fig. 7.1 is contingent upon a logistic distribution of the additive noise.

Now let the temperature ($T = 1/\beta$) be a function of time (t) with the intention of *cooling* the Boltzmann machine in a gradual way, eventually reaching $T \approx 0$ where the "irreversible" process (of convergence to stable states) takes over. If it were possible to merely insert a time-dependent temperature T(t) into the formula for the stationary distribution, as:

$$p(x) = \frac{e^{-\beta(t)H(x)}}{Z(t)}, \qquad Z(t) = \sum_{x} e^{-\beta(t)H(x)}, \tag{1}$$

then the network state would, as $T(t) \to 0$, (*almost surely*) converge to the **deepest** energy minimum.

1. IRREVERSIBLE ND.

AT $T = 0$, $\underline{X}(t) \xrightarrow{P.} \underline{x}^0 \in \{$ LOCAL MINIMA OF $H(\underline{x}) \}$.

2. EQUILIBRIUM STATISTICAL ND.

AT $1/T = \beta < \infty$, $\underline{X}(t) \xrightarrow{d.}$ Boltz., ie., $p(\underline{x}) = Z^{-1} \exp[-\beta H(\underline{x})]$.

3. COMPUTATIONAL ND.

FOR $1/T = \beta^*(t)$, $\beta^* \to \infty$ AS $t \to \infty$, $X(t) \xrightarrow{a.s.} \underline{x}^0$,

WHERE $\underline{x}^0$: $H(\underline{x}^0) = \inf_{\underline{x}} H(\underline{x})$.

Figure 7.1 Fundamental principles of collective computation.

Problem 7.1: A symmetric network of binary neurons has a state $\mathbf{x}^*$ of inimum energy, $H(\mathbf{x}^*) = -a$. Every other state has energy no less than $-a+b$: $(\mathbf{x}) \geq b-a$ for all $\mathbf{x} \neq \mathbf{x}^*$, (a and b positive).

(a) Derive an upper bound on the partition function at temperature $T = 1/\beta$.

(b) Derive a lower bound on the (equilibrium) modal probability $P(\mathbf{x}^*) = p^*$.

(c) For fixed, positive, finite N, a, and b, evaluate the lower bound on p^* in the limit as β approaches infinity.

Solution:

(a) The partition function is:

$$Z = \sum_x e^{-\beta H(x)} \leq e^{\beta a} + (2^N - 1)e^{\beta(a-b)}$$

(b) $p^* = e^{\beta a}/Z \geq 1/(1 + 2^N e^{-\beta b})$.

(c) $\lim_{T \to 0}(1 + 2^N e^{-\beta b})^{-1} = 1$; therefore $p^* \to 1$ as T goes to zero. No matter how small the (non-zero) energy gap between $\mathbf{x}^*$

and its closest competitor, the probability of $\mathbf{x}^*$ will alwa converge to unity (for finite N) as the temperature goes zero.

This claim, however, depends on the validity of equation (1) at eve step, t, of the cooling process. The derivation of the Boltzmann machine in t last chapter relied on the argument that the sequence of network states {**X**($t \geq 0$} is a homogeneous, irreducible, aperiodic Markov chain with a stationa distribution like (1). Homogeneity was assured by the time-invariance of t parameters; and the time-dependence of the temperature immediately casts t theorem in doubt.

The Third Principle of fig. 7.1 asserts the existence of a cooli schedule so gradual that convergence of the network state to a global ener minimum is almost sure. In the context of collective computation networks f constrained optimization (chapter 5) this means cooling to a final state that is t best possible solution (or one of the co-optimal solutions) to the proble represented by the network.

Thermodynamic metaphors

Atoms of a molten metal, when cooled to a freezing temperature, w tend to assume relative positions in a lattice in such a way as to minimize t potential energy of their collective interaction. Because the number of atoms so great, the final state will most likely correspond to only a local minimum a not a global one. As investigators in prehistoric times found out, a solidifi metal may be reheated and cooled slowly to overcome the brittleness that (as now know) characterizes the shallower energy minima. In metallurgy th process is called annealing.

Simulated annealing is a way to find a global extremum of a functi that has many local extrema and may not be smooth (Bohachevsky et al, 198 The method is a biased random walk that samples the objective function in t space of the independent variables.

Simulated annealing with "fast computing machines" traces its roots the work of Metropolis et al. (1953), who introduced the following dynami assumptions:

(a) If x_1, x_2,... are the positions of particles in a lattice, mo each particle in succession so that:

$$x_1 \rightarrow x_1 + b\xi_1$$
$$x_2 \rightarrow x_2 + b\xi_2$$
$$\ldots \text{ etc.}$$

where b is the maximum allowed displacement and $\xi_1, \xi_2, \ldots$ are random numbers on the unit interval.

(b) Calculate the change in energy (ΔH) of the system.

(c) If $\Delta H \leq 0$, let the particle keep its updated position; but if $\Delta H > 0$, retain the new position with probability $p = \exp(-\Delta H/kT)$ where k is Boltzmann's constant and T is the absolute temperature. Metropolis, Rosenbluth, Rosenbluth, Teller, and Teller (1953) showed that this algorithm "chooses configurations with probability $\exp(-H/kT)$."

Problem 7.2: Define the Boltzmann machine in terms of the Metropolis algorithm. Assume McCulloch-Pitts neurons.

Solution: Let $\xi(t)$ be equiprobably zero or one and let $\{K(t), t=1,2,\ldots\}$ be i.i.d. uniformly on $\{1,\ldots,N\}$. For $i=K(t)$ let:

$$\chi_i = x_i(t) \oplus \xi(t)$$

where $\oplus$ denotes modulo-2 addition: $0 \oplus 1 = 1$ and $1 \oplus 1 = 0$. Calculate the energy gap:

$$\Delta H = H(\chi_i;\mathbf{x}(t)) - H(\mathbf{x}(t)).$$

Where $(\chi_i, \mathbf{x}(t))$ means $x_i(t)$ is to be replaced by χ_i. If $\Delta H \leq 0$, let $x_i(t+1) = \chi_i$. Otherwise let:

$$X_i(t+1) = \begin{cases} \chi_i \text{ with probability } p = e^{-\Delta H/T} \\ X_i(t) \text{ with probability } 1-p \end{cases} \quad \textbf{(2)}$$

This procedure is the same as iterating the Glauber dynamical equation:

$$X_i\ (t+1) = \begin{cases} \mathcal{H}\ [\sum_j W_{ij}\ X_j(t)\ +\ Y(t)] & \textit{if } i=K(t) \\ X_i\ (t) & \textit{otherwise} \end{cases}$$

with logistic noise: $\Pr[Y(t) \leq y] = 1/(1 + e^{-y/T})$. Both procedures lead to:

$$X_i\ (t+1) = \begin{cases} 1 \textit{ with prob.} & \frac{1}{2}\ [1 + \tanh \frac{\frac{1}{2}\sum_j W_{ij}\ X_j(t)}{T}] \\ 0 \textit{ with prob.} & \frac{1}{2}\ [1 - \tanh \frac{\frac{1}{2}\sum_j W_{ij}\ X_j(t)}{T}] \end{cases}$$

and $\frac{1}{2}\ [1 + \tanh(y/2T)] = 1/[1 + \exp(-y/T)]$.

To see this more clearly, suppose that the transition from $x_i = 0$ to $x_i = 1$ goes downhill in energy. Then, if $x_i(t) = 0$, the occurrence of $\xi(t) = 1$ (with probability ½) sends the network state downhill; and $\xi(t) = 0$ (with probability ½) leaves everything the same. Similarly, if $x_i(t) = 1$ and $\xi(t) = 0$, then $\Delta H = 0$ and no change is made. But, if $x_i(t) = 1$ and $\xi(t) = 1$, we calculate the gap ΔH and move uphill with probability:

$$\Pr[x_i(t+1) = 0 \mid x_i(t) = 1) = p(\Delta H)\ \Pr[\xi(t) = 1] = p/2$$

as instructed by equation (2).

Now let the stationary probabilities for the ith unit be denoted $\pi_0 = \Pr(x_i = 0)$ and $\pi_1 = \Pr(x_i = 1)$. The diagram leads to a pair of simultaneous equations, one of which is:

$$\pi_0 = \tfrac{1}{2}\ \pi_0 + \tfrac{1}{2}\ p\ \pi_1$$

Since $\pi_0 + \pi_1 = 1$, the result is $\pi_1 = 1/(1+p)$. Substituting $p = e^{-\Delta H/T}$ gives:

$$\pi_1 = 1/(1 + e^{-\Delta H/T}) = \tfrac{1}{2}[1 + \tanh(\Delta H/2T)]$$

where the second equality follows from the solution to problem 1.1. Writing $x_i(t+1) = 1$ with probability π_1 merely takes the limit as t approached infinity.

The reader will discern an inconsistency — really a multiplicity — in the historical synopsis above, since we have discussed annealing schedules of three distinctly different types. With reference to fig. 7.2 we can:

(a) alternately heat and freeze the system (like the Bronze Age smith);

(b) immerse the system in a heat bath at constant temperature (like Metropolis); or

(c) cool the system so gradually that the optimum state is reached in the limit. Option (c) is usually what researchers talk about today under the monicker of "simulated annealing".

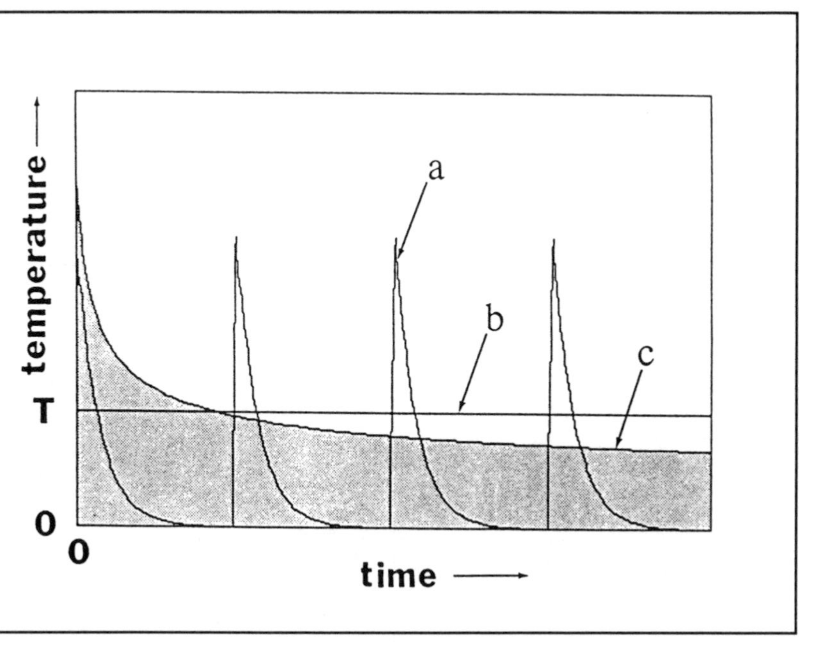

Figure 7.2 Temperature schedules for finding improved solutions.

Thermal Cycling

Figure 7.3 shows a (successful) attempt to solve a five-city Euclidean SP by thermal cycling of a collective computation network. (The problem was osed earlier in fig. 5.2.) A "heat pulse" was applied to the network after its onvergence at zero temperature to a clearly suboptimal solution of tour length 675 miles. The pulse duration is about twenty sweeps, each sweep consisting f $N = 5^2$ iterations of equation (2). The mileage is divided by 50 (miles) in e program that performs this exercise. The excitatory offset and inhibitory rength are $B=15$ and $C=30$ dimensionless units, respectively; so the weakest (nonzero) excitatory link in the net is B - D_{max} ≈ 15 - (600mi)/(50mi) = dimensionless units. That the "heat pulse" peaks at T = 18 in the first cycl means that the r.m.s. intensity of the additive noise is on the same order as th strength of the strongest excitatory connections. The second cycle uses a puls with two peaks, as the control of the temperature is via keyboard by programmer who aborted the cooling process when it appeared to be leading t a longer tour. This kind of human-computer interaction may be akin t metallurgical practice circa 1200 B.C. The obvious task is to apply scientifi methodology to the determination of universally effective annealing schedules

Geman and Geman (1984) proposed an annealing schedule:

$$\beta(t) = \beta_o \ln(1 + t) \quad \textbf{(3a}$$

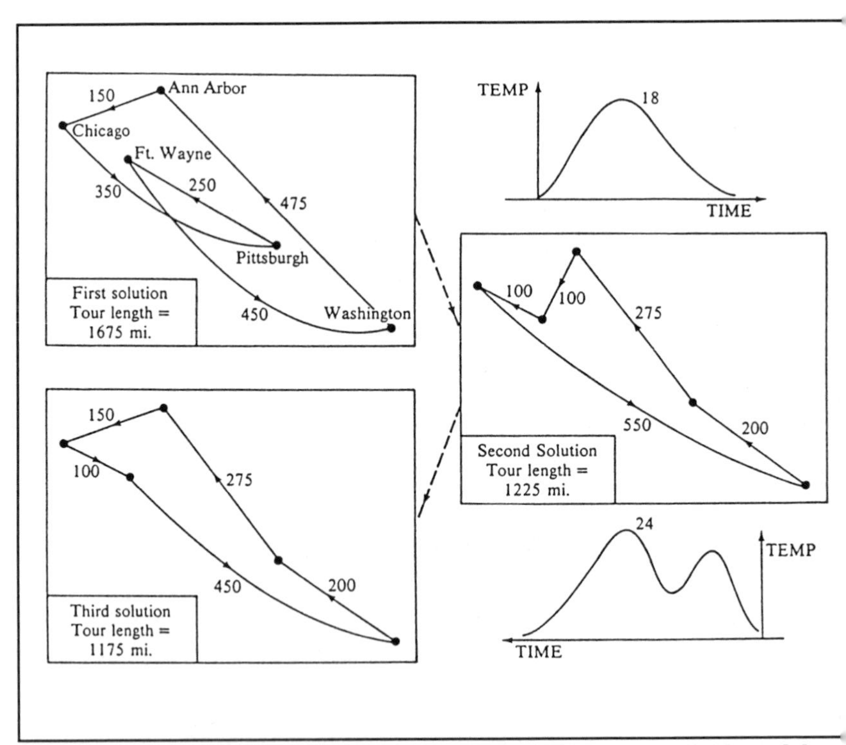

Figure 7.3 Solutions to a Five City TSP using the Hopfield model.

and offered a proof of its effectiveness in the form of the Third Principle (fig. 7.1). They prescribed:

$$1/\beta_0 = T_0 = N(\Delta H)^* , \qquad \textbf{(4)}$$

where N is the network size and $(\Delta H)^* = \max_x H(\mathbf{x}) - \min_x H(\mathbf{x})$ is the full vertical span of the energy landscape. Their computational experience in image restoration suggested that this schedule could be accelerated by starting from a much lower temperature and counting time in sweeps (see page 80) instead of "site visits" (or iterations of equation (2)):

$$\beta(t) = \tilde{\beta}_0 \ln(1 + t/N), \quad \tilde{\beta}_0 >> [N(\Delta H)^*]^{-1} . \qquad \textbf{(3')}$$

Problem 7.3: With reference to problem 7.1(b), how many steps does it take to attain $p^* \geq 1/2$ when the cooling schedule is given by equation (3a)? How many sweeps using (3′)?

Solution: Substitute $\beta(t)$ for β in the expression for $p^* \geq 1/[1 + 2^N \exp(-\beta_0 \ln(1+t))] \approx 1/(1 + 2^N t^{-\beta_0 b})$ for $t >> 1$. Setting $p^* = 1/2$, the result is $t^{\beta_0 b} = 2^N$, ie., $t = 2^{N/[\beta_0 b]}$ steps. Using equation (4) for β_0, this is t, where:

$$\log_2(t) = N^2(\Delta H)^*/b \qquad \textbf{(5a)}$$

to guarantee at least 50% probability of convergence to the global optimum. On the other hand, if equation (3′) is used:

$$t/N = 2^{N/[\beta_0 b]} \text{ sweeps.} \qquad \textbf{(5b)}$$

Even if it is possible to choose the initial temperature so that the difference between it and $(\Delta H)^*$ will be much smaller than b, the number of sweeps required to find the optimum (with probability 1/2) is about 2^N. *The state space can be searched exhaustively with the same amount of computation!*

Since the logarithmic cooling schedule of Geman and Geman is so slow as to cast doubt on the practicability of simulated annealing in digital simulation models of even moderately large networks, some researchers have sought to accelerate the almost sure convergence to global optima by means of faster cooling schedules and a variety of intriguing interpretations of the meaning of temperature in the stochastic model. This line of investigation is exemplified by Szu and Hartley (1987), Akiyama et al. (1989) and Matsuba (1989). Despite the encouraging computational experience of these and other investigators, the mathematical foundations of "fast simulated annealing" still need considerable clarification. In view of the Gibbs-Markov equivalence, asserted in the preceding chapter, we must resist the idea that the distribution of global energy can be profoundly altered by deviating the noise amplitude distribution from the canonical (logistic) form. Fig. 7.4 shows the empirical variation of modal probability with asymptotic temperature in the system of two neurons studied in problem 6.5. Appendix B develops the formula:

$$T_\infty = 1/[4dF(y)/dy]_{y=0}$$

for the asymptotic temperature of a network in which each unit is driven by an independent white noise source with zero mean and S-shaped amplitude distribution function F(y). As the asymptotic temperature grows large, the network behaves more nearly as a Boltzmann machine; but at sufficiently low temperatures, the Boltzmann distribution at temperature T_∞ is only a crude approximation to the stationary distribution.

If numerical studies are to reveal anything about the theoretical foundations of fast simulated annealing, they must clarify the evident trade-off of computational speed versus solution quality. On the one extreme, the logarithmic cooling schedule guarantees optimality but works hardly faster than exhaustive search. On the other extreme we have thermal cycling between an elevated temperature (which effectively picks an initial state at random) and zero temperature (which leads to stable states in a few sweeps). This latter approach returns us to the decidedly poor performance observed in chapter 5, with states being picked at random from the solution space, giving rise to a uniform distribution of solution quality. Between these two extremes there must be a "happy medium" in the form a cooling schedule that is fast enough to compute but gradual enough to find truly good (although suboptimal) solutions.

Numerical Experiments

Let the meaning of "cycle" be revised in the context of a computational procedure like the following, in which the weights are **W**(.,.), the states are **x**(.), and the units are indexed by I or J:

```
...
200 FOR CYCLE = 1 TO 10
210   TEMP = T0*(EXP(-.46*CYCLE))-.01)
220   FOR TIME = 1 TO ALPHA*N
230     I = INT(RND*N))
240     U = TEMP*LOG(1/RND - 1)
```

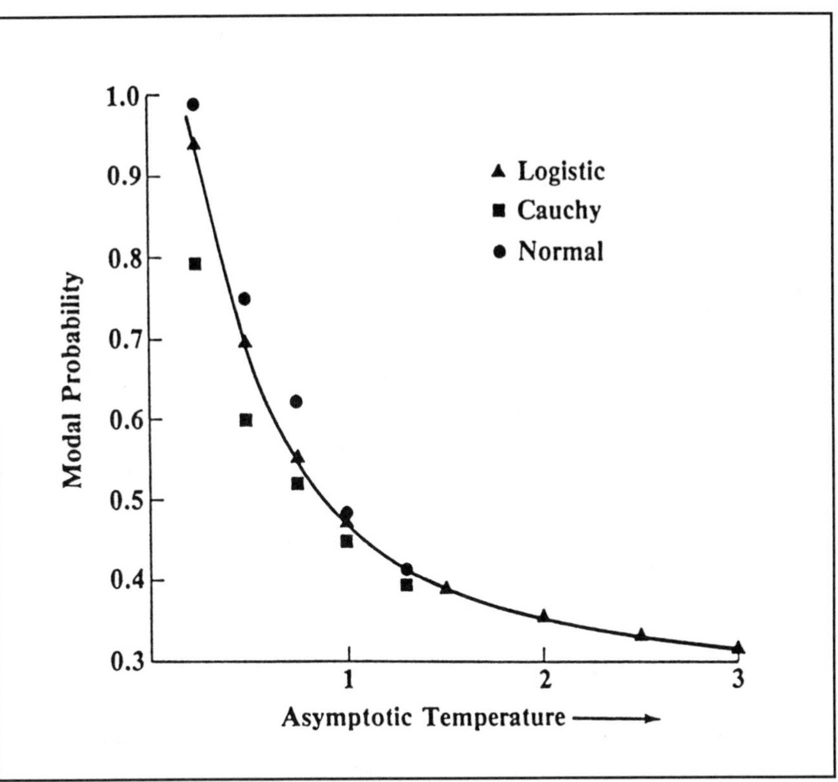

Figure 7.4 Modal probability versus temperature (T_∞) for neuron pair with three kinds of noise.

```
250     FOR J = 1 TO N
260       U = U + W(I,J)*X(J)
270       NEXT J
280     IF U>0 THEN X(I)=1 ELSE X(I)=0
290     NEXT TIME
300   NEXT CYCLE
...
```

The end product of this procedure is a network state x(1)...x(N). If line 300 is followed by several more sweeps of the kind initiated in 220 — but with TEMP equal to zero — the network will reach a stable state. Moreover, if ALPHA > 10 *sweeps per cycle*, a single cycle will probably be sufficient to obtain convergence.

A smooth approximation to the stepwise cooling schedule just described is:

$$(\text{TEMP} =) \ \ T(t) = T_o(e^{-.46\alpha t} - .01), \ \ 0 \leq \alpha t \leq 10,$$

where t counts the number of sweeps. Each cycle involves α sweeps of the net. Thus $T = 0$ after $10\alpha N$ iterations of equation (2). Letting α vary from .5 to 8, this schedule was used to cool a sixteen-unit TSP net. The Euclidean problem was that of four cities at the corners of the unit square (fig. 5.7). With the excitatory offset (B) and inhibitory strength (C) as indicated in fig. 7.5, there are six stable states when a strong, positive bias is applied to the unit representing the origin of the tour. (Note: the initial temperature is 2C.) Of the $(M-1)! = 3! = 6$ tours, two are represented by the square and four by the hourglass (two possible orientations and two directions). Thus the probability of obtaining an optimal tour is $2/6 = 1/3$ when solutions are picked at random. When $\alpha < 3$ sweeps per cycle, the observed performance is actually worse than this; but between 3 and 5 sweeps/cycle the performance rises to about 80% correct. Unfortunately, this level seems to be a plateau; and further prolonging the cooling process has no apparent advantage.

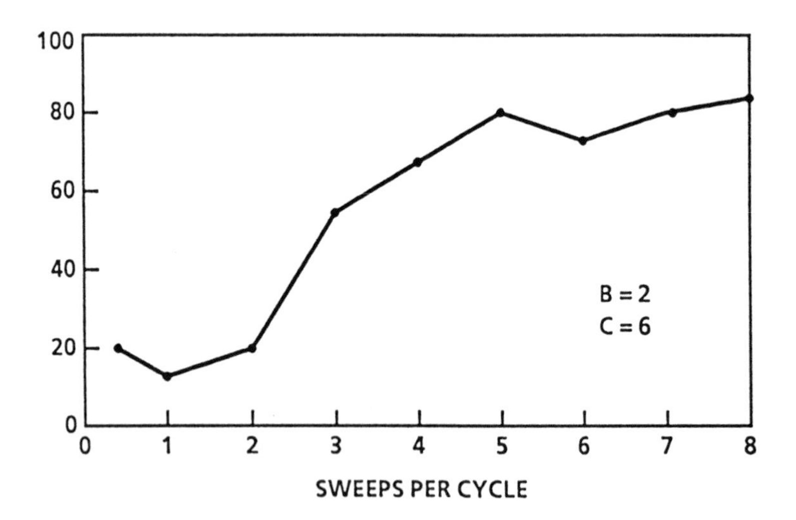

Figure 7.5 Percent correct vs. sweeps per cycle using exponential cooling schedule with an E-folding time of 2.17 c Each plotted point represents 15 runs of the 4-city TSP in which the unit square is the optimal solution.

Fig. 7.6 compares the exponential schedule to a reciprocal time schedule which likewise stops at zero after ten cycles. The number of sweeps is t, the number of sweeps per cycle is α, and $\alpha t \leq 10$. Fig. 7.7 shows computed results for both schedules applied to the five-city, non-Euclidean TSP (Figure 5.10). (Note: the initial temperature is 2C where C is the strength of the inhibitory lengths between units sharing the same row or column.) The vertical scale is the cost (in $) of the tour by air averaged over twenty trials at each value of parameter α. Here again we see a marked improvement as α rises through the range of 2 to 8, especially for the reciprocal time schedule. But the performance level reaches a plateau which extends beyond sixty four sweeps/cycle. The dotted line indicates the optimum performance level.

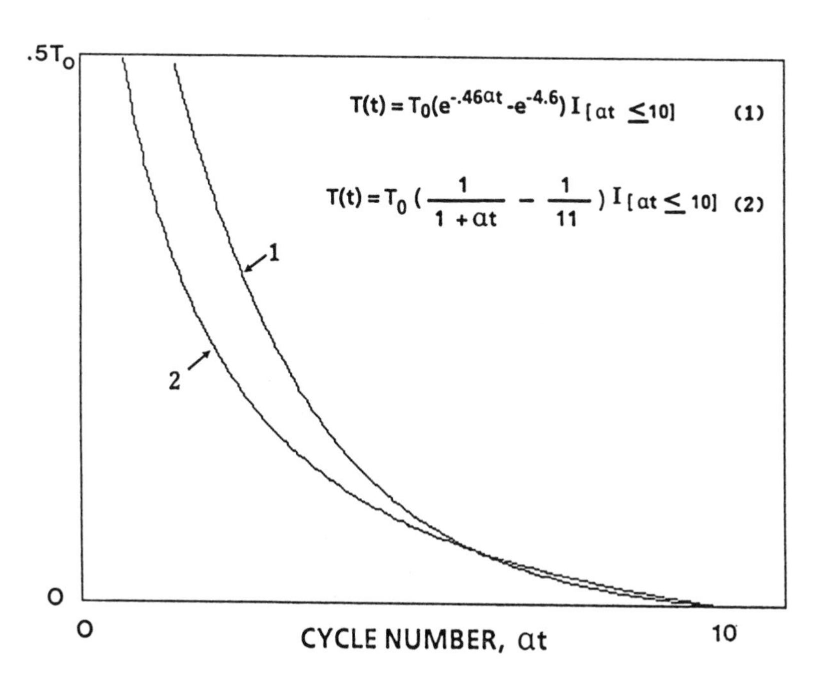

Figure 7.6 Temperature vs. cycle number for two cooling schedules: the exponential and the reciprocal time.

From these experiments with "toy problems" it might be possible to infer a lesson of broad generality: A little gradualness in the cooling of the network gives a significant gain in solution quality. However, to attain optimality with a high level of confidence, the cooling must be exceedingly

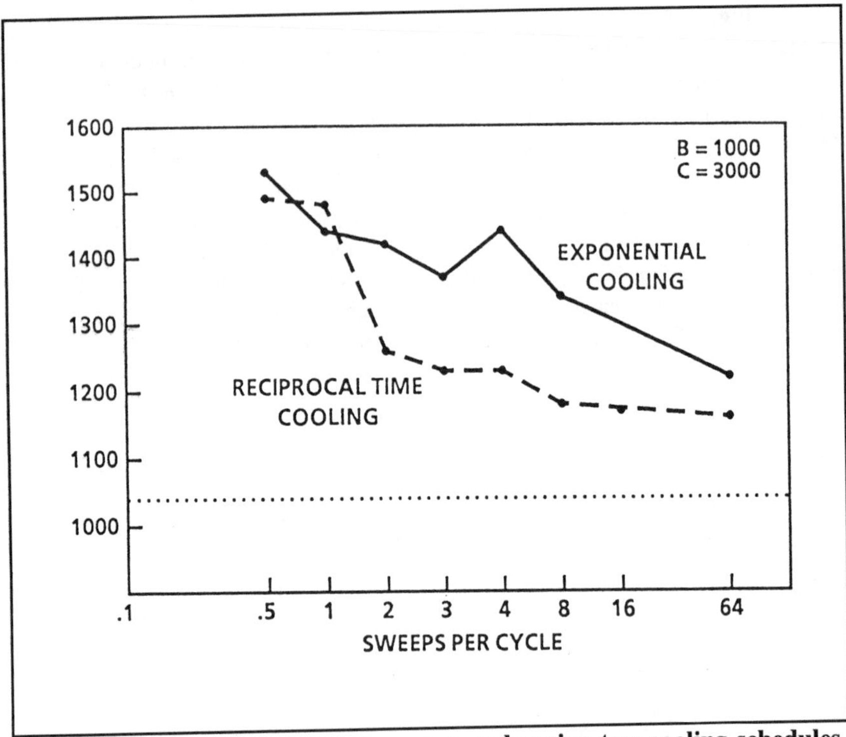

Figure 7.7 Average cost vs. sweeps per cycle using two cooling schedules, the exponential (solid) and reciprocal time (dashed). Cost of the tour (by air) is averaged over 20 runs of the 5-city TSP.

gradual. This heuristic notion raises a question of computational expediency: Is it better to cool very gradually and take the one result, or to cool less gradually several times in succession and take the best of several solutions? The next problem tries to illustrate the question. Since its quantitative assumptions are so conjectural, however, the result should not be construed as a definite answer to the general case.

Problem 7.4: The Amalgamated Brush Company purchases ten seconds of CPU time from a telecomputing service to minimize the total airfare of a ten-city tour by its newest salesperson. Each second of CPU time executes 100 sweeps of a 100-unit TSP network with the appropriate (symmetric) weight matrix. The software uses a cooling schedule that drops to zero temperature in K sweeps (after which convergence is comparatively instantaneous). Suppose the cost of the solution is uniformly distributed between the absolute minimum cost, $\$_{min}$, and $\$_{max} = \$10000L + \$_{min}$, $L = \ln(2)/\ln(K+2)$. Thus the excess cost $\$_x = S_{max} - S_{min}$ goes to zero as K approaches infinity. Should the CPU time be used to obtain one solution at K=1000 sweeps, or ten solutions at K=100 sweeps each?

Solution: Take one solution after cooling over 1000 sweeps. If M is the number of solutions obtained, then KM = 1000. The distribution of the cost of the *best of M* solutions, is:

$$\$^* = \min[\$_x(1),\ldots,\$_x(M)],$$

$$\Pr(\$^*/\$10000 \leq y) = (y/L)^M,\ 0 \leq y \leq L,$$

when the solutions are independent. The expected value is:

$$E\$^* = \$10000 \int_0^L [\ 1 - (y/L)^M\ ]\ dy$$

$$= \$10000L[1 - \frac{1}{(M+1)}]$$

$$= \$10000\ [\frac{.69}{\ln(1000\ /M + 2)}\ ]\ [\ 1 - \frac{1}{(M+1)}]$$

after using the above value for L and substituting 1000/M for K. Then if 1000 sweeps give one solution, its expected cost is:

$$\$10000[.69/\ln(1002)][1/2] \approx \$500$$

more than the optimum. Taking the best of ten, $E\$^* \approx \$1960$. Of course, if the ten consecutive solutions are **not** independent, the sequence $\{\$_x(m), m=1,2,\ldots\}$ exhibiting a decreasing trend as shallower minima are abandoned in favor of deeper ones, the result could be quite different.

The Isothermal Boltzmann Machine

The isothermal Boltzmann machine has received scant attention as a computational device, probably because its sequence of states cannot be treated as output without some kind of back-end statistical processing. After several sweeps at a moderate, constant temperature, the probability of having a stable state should be approximately:

$$Pr[X \in \{x^0\}] = \frac{\sum_{x \in \{x^0\}} \exp[-H(x)/T]}{\sum_{x \in S} \exp[-H(x)/T]}$$

with S being the state space. Because $\#\{\mathbf{x}^o\}/2^N$ is so small, this probability will tend to zero, at least above some critical temperature, as the number of units increases.

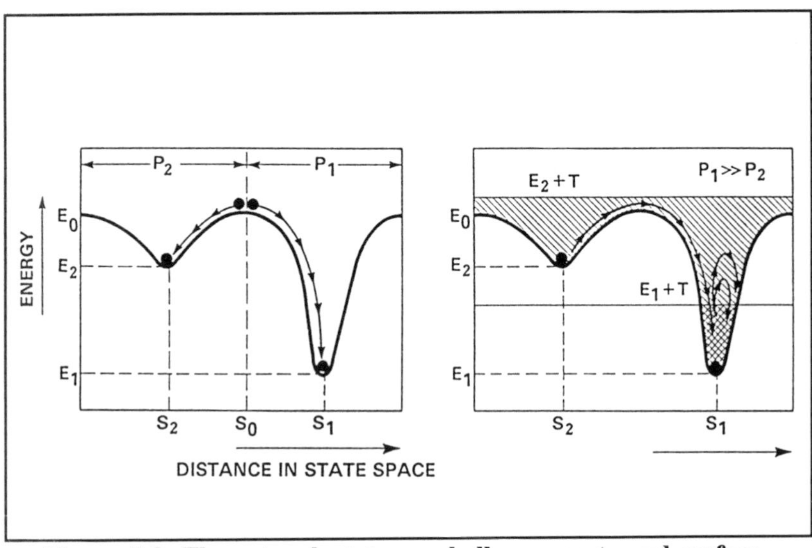

Figure 7.8 The network state as a ball on a contoured surface.

Figure 7.8 sketches the specious likeness of the Boltzmann machine to a pinball on a contoured surface. On the left side, corresponding to zero temperature, the pinball rolls to the bottom of the well in which its initial position resides; and the likelihood of settling into a given well, when the initial position is uniformly distributed, is determined by the longitudinal extent of the wells (or basins of attraction). On the right, this mechanical analog is placed on a vibrating table. If the intensity of the vibrations is properly adjusted, one may expect the ball to quickly be bounced out of the shallower well and into the deeper one — from which it will not escape for a very long time if the intensity of the vibrations (corresponding to temperature) stays the same.

The problem is that, even after the state of the isothermal Boltzmann machine has "fallen into the deepest well", it continues to wander aimlessly in the vicinity of the optimum. What kind of data processor can be employed to "decode" this random activity? It would seem straightforward enough to program a general-purpose computer to read the states of the Boltzmann machine and evaluate them as candidate solutions to the problem at hand. Some kind of stopping rule will need to be devised. But if one should insist on having the neurons declare their consensus more directly — so that users who are uncomfortable with statistical tricks can "put their fingers on the answer" — the sketch in fig. 7.9 might be worth examining. Here we have two collective computation networks, now subnets, which are identical in connectivity but different in their temperatures and sources of bias. The input subnet may be biased by the parameters of an optimization problem. It is in equilibrium at a temperature T_1 sufficiently high to produce a state sequence in which the valid solution states are greatly outnumbered. The second subnet is at temperature T_2 close to zero; and excitatory connections of strength b feed forward from the first subnet to the second. The process $\{\mathbf{x}^{(1)}(t), t \geq 0\}$ thus constitutes the input to the second subnet whose state sequence $\{\mathbf{x}^{(2)}, t \geq 0\}$ may be dominated by stable states for a suitable choice of b. In this way the designer might motivate the second subnet to be attracted to stable states with relative frequencies determined by the *depths* of their energy basins. The reader who is intrigued by this possibility can refer to appendix D for details and some results of numerical experiments with subnets coupled in the manner of fig. 7.9.

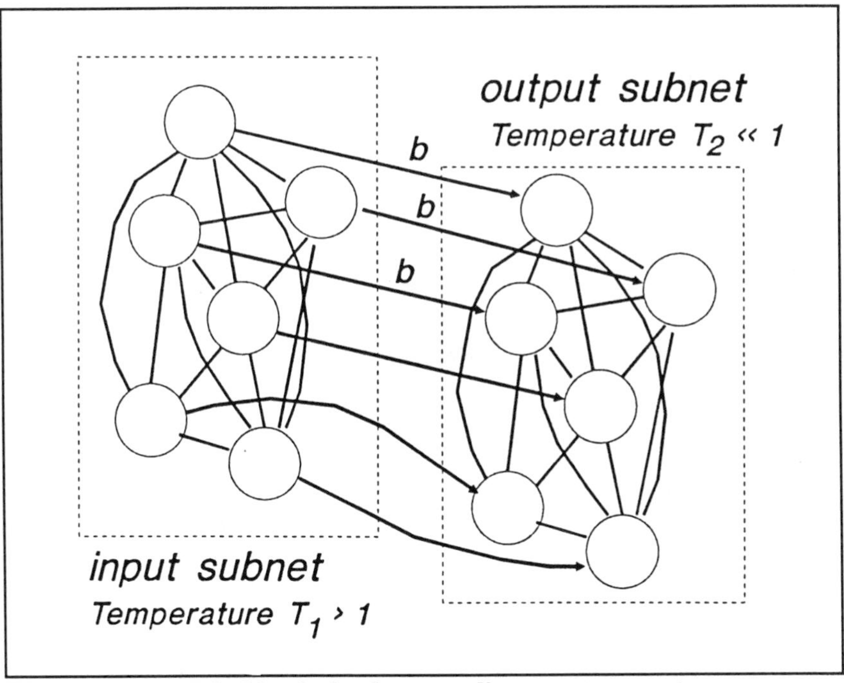

Figure 7.9 Identical subnets at disparate temperatures are linked by one-way excitatory connections.

8. Data Representation and Pattern Storage in Content-Addressable Memory

James Anderson, who is well known for prescient observations regarding connectionist models, has been heard to state two generalizations that are pertinent to this chapter. The first is that data representation is crucially important to the success of any neural network design or training exercise. (The right data representation may lead to good results in spite of mistakes further downstream in the design process; and the wrong representation will defeat the designer who makes no other mistakes.) The second is that a connectionist model is essentially a way of capturing statistical correlations among suitably represented data items. This theme will be studied before returning to the first.

After a long but essential digression we again cast the collective computation net in the role of a content-addressable memory (CAM), which it fills by converging to stable states at energy minima — subject to the requirement that there be a way to place these minima at the state space coordinates of desired "patterns."

The Outer Product Rule

Recalling chapter 1 (equation 5), Hopfield used the outer product rule to satisfy the requirement just noted. If $\mathbf{x}^1,\ldots,\mathbf{x}^M$ are N-bit patterns (vectors), $\mathbf{x}^m = (x_1^m,\ldots,x_N^m)$, $x_i^m \in \{0,1\}$, invoke the step-to-sign transformation $s = 2x-1$. (See p. 3.) The M x N pattern matrix:

$$S = \begin{bmatrix} s_1^1 & \cdots & s_N^1 \\ \cdot & & \cdot \\ \cdot & & \cdot \\ \cdot & & \cdot \\ s_1^M & \cdots & s_N^M \end{bmatrix} = \begin{bmatrix} s^1 \\ \cdot \\ \cdot \\ \cdot \\ s^M \end{bmatrix}$$

gives rise to an NxN weight matrix $i \neq j$. (The normalizing factor 1/M is optional.) As Van Hemmen et al. (1988) have pointed out, this equation (1) is the same as the inner (dot) product rule $W_{ij} = \mathbf{s}_i \bullet \mathbf{s}_j / M$ in terms of the columns of **S**.

$$\mathbf{W}:\ W_{ij} = (1/M)\sum_{m=1}^{M} s_i^m s_j^m = (1/M)\sum_{m=1}^{M} (2x_i^m - 1)(2x_j^m - 1) \quad , \qquad \textbf{(1)}$$

The Hinton-Sejnowski Formula

A different storage prescription, due to Hinton and Sejnowski (1983), is written by Rumelhart et al. (1987) as:

$$W_{ij} = -\ln \frac{p(x_i = 0 \ \&\ x_j = 1)\ p(x_i = 1 \ \&\ x_j = 0)}{p(x_i = 1 \ \&\ x_j = 1)\ p(x_i = 0 \ \&\ x_j = 0)} \qquad \textbf{(2a)}$$

$i \neq j$, in which the probability, p, is calculated from the frequency of occurrence in the pattern set, eg.:

$p(x_i = 0 \ \&\ x_j = 1)$ = the fraction of patterns for which unit i is OFF and unit j is ON
= the probability of having $x_i^m = 0$ and $x_j^m = 1$ when $\mathbf{x}^m$ is equiprobably any row of **S**,

when McCulloch-Pitts neurons are assumed. (If the patterns are given in terms of signs, invoke the sign-to-step transformation [Problem 1.1]) before applying the formula.) Following Rumelhart (1987), four aspects of equation (2a) are noted:

(1) If two units tend to be ON or OFF <u>together</u> in the pattern set, then the weight will be a large <u>positive</u> value.

(2) If two units assume <u>different</u> states much more often than they agree, then the weight on the link between them has a large <u>negative</u> value.

(3) If two units behave <u>independently</u> in the pattern set, so that $p(x_i = a \ \&\ x_j = b) = p(x_i = a)p(x_j = b)$, then the weight is <u>zero</u>.

(4) The formula assures <u>symmetry</u>, as $W_{ij} = W_{ji}$.

In addition, each unit is given a bias (constant input):

$$z_i = \ln[p(x_i=1)/p(x_i=0)]; \quad \textbf{(2b}$$

so if the unit is usually OFF it has a negative bias, etc.

The Pattern Correlation Matrix

A slightly more compact notation can be used to redefine th probabilities in the Hinton-Sejnowski formula. Let:

$$\mathbf{c}: \quad c_{ij} = (1/M)\sum_{m=1}^{M} x_i^m x_j^m \quad \textit{for all i and j} \quad \textbf{(3)}$$

be called the pattern correlation matrix. Fig. 8.1 illustrates the computation c **c** for six nine-bit patterns. Strictly speaking, we should note that the covarianc is:

$$\mathrm{Cov}(x_i,x_j) = E(x_i x_j) - (Ex_i)(Ex_j) \ ,$$

given that $\Pr[\mathbf{x} = \mathbf{x}^m] = 1/M$ for every $m \in \{1,\ldots,M\}$; and the correlatio coefficient follows as:

$$\rho_{ij} = \frac{Cov(X_i,X_j)}{[\ Var(X_i)\ Var(X_j)\ ]^{1/2}} = \frac{c_{ij} - c_{ii}c_{jj}}{[c_{ii}c_{jj}(1-c_{ii})(1-c_{jj})]^{1/2}} \quad \textbf{(4)}$$

Perhaps this misnomer can be forgiven if we are consistent in our abuse c statistical terminology! Such abuse will make it possible to illustrate the liter accuracy of the second point attributed to Anderson above: Both storag prescriptions (1) and (2) merely capture the statistical correlations among th components of suitably represented data items.

Regarding the joint probabilities in the Hinton-Sejnowski formula:

$$p(x_i=1 \ \&\ x_j=1) = c_{ij} \qquad p(x_i=0 \ \&\ x_j=0) = 1 + c_{ij} - c_{ii} - c_{jj}$$
$$p(x_i=1 \ \&\ x_j=0) = c_{ii} - c_{ij} \qquad p(x_i=0 \ \&\ x_j=1) = c_{jj} - c_{ij}. \quad \textbf{(5}$$

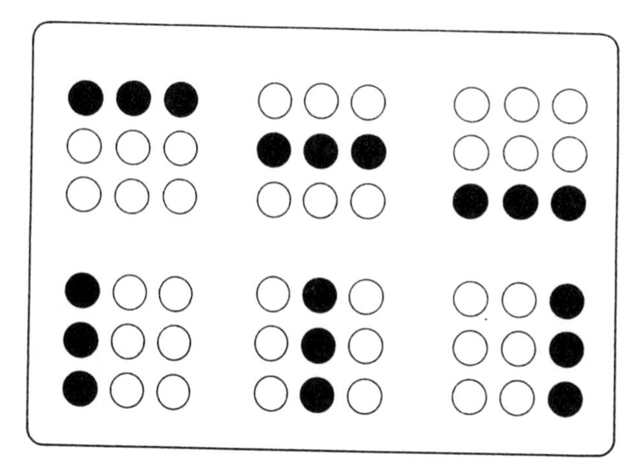

Pattern Correlation Matrix, **c**:

$$c_{ii} = 2/6 = 1/3 \quad \text{for every } i$$

$$c_{ij} = \begin{cases} 1/6 & \text{if } i \text{ and } j \text{ share same row or column} \\ 0 & \text{otherwise} \end{cases}$$

Figure 8.1 Six patterns to store in a nine-bit C.A.M.

For instance, the last (lower-right) identity is derived as

$$p(x_i{=}0 \;\&\; x_j{=}1) \;=\; E[(1-X_i)X_j] \;=\; (1/M)\sum_{m=1}^{M}(x_j^m - x_i^m x_j^m) \quad .$$

Note that the sum of the right sides of equations (5) is unity. Alternatively, the lower left may be derived by a simple subtraction. Count the number of times x_i is ON, (Mc_{ii}) and count the number of times both x_i and x_j are ON (Mc_{ij}). The difference is the number of times when x_i is ON and x_j is not. Divide by the total number of patterns to obtain the frequency of occurrence and thence the probability.

Hence equation (2a) becomes:

$$W_{ij} = \ln \frac{c_{ij}\ (1+c_{ij}-c_{ii}-c_{jj})}{(c_{ii}-c_{ij})\ (c_{jj}-c_{ij})} \qquad (i \neq j). \tag{6}$$

The outer product rule (1) also reduces to a function of the components of **c**:

$$\frac{\sum_m (2x_i^m-1)(2x_j^m-1)}{M} = \frac{\sum_m (4x_i^m x_j^m - 2x_i^m - 2x_j^m + 1)}{M}. \tag{7}$$

$$= 4c_{ij} - 2c_{ii} - 2c_{jj} + 1 = W_{ij}$$

This obvious but noteworthy point is repeated in fig. 8.2 using q (instead of M) for #{patterns}. Fig. 8.3 illustrates the use of both storage prescriptions (6) and (7) for the same pattern set as fig. 8.1.

The Hinton-Sejnowski formula, unlike Hopfield's prescription, is vexed by singularities when:

$[c_{ij} = 0] \Rightarrow$	units i & j <u>never</u> turn ON to the same pattern, or
$[1+c_{ij}-c_{ii}-c_{jj}=0] \Rightarrow$	units i & j <u>always</u> turn ON to the same patterns, or
$[c_{ii}-c_{ij}=0] \Rightarrow$	unit i can turn ON to a pattern <u>only</u> when unit j does.

Appendix C, which offers a derivation of the Hinton-Sejnowski formula somewhat different from the originators', suggests optional treatments of these contingencies (and provides several examples of the formula's use). Without much ado, let us propose the quick fix:

$$W_{ij} = \ln\{[c_{ij}'(1+c_{ij}'-c_{ii}-c_{jj})]/[(c_{ii}-c_{ij}'')(c_{jj}-c_{ij}'')]\} \tag{6'}$$

$$(2x_i - 1)(2x_j - 1) = 4x_ix_j - 2x_i - 2x_j + 1$$

$$\Rightarrow$$

$$\frac{1}{q}\sum_{k=1}^{q}(2x_i^k - 1)(2x_j^k - 1) = 4c_{ij} - 2c_{ii} - 2c_{jj} + 1$$

$$\text{where } c_{ij} = \frac{1}{q}\sum_{k=1}^{q} x_i^k x_j^k \ \forall i,j$$

$$\text{NOTE: } c_{ij} = \#\{k : x_i^k = x_j^k = 1\}/\#\{\text{patterns}\}$$

Figure 8.2 Pattern correlation matrix.

with the definitions $c_{ij}' = c_{ij} + \epsilon$ and $c_{ij}'' = c_{ij} - \epsilon$ for a suitably small ϵ ($<1/M$). This fix is tantamount to the assertion that, if the pattern set were (somehow) enumerated a little further, at least one pattern would eliminate any instance of the singularities noted. Its effect is to place upper and lower bounds proportional to $\pm\ln(\epsilon)$ respectively on the strong inhibitory (or excitatory) links between units that represent contradictory (or redundant) hypotheses.

Problem 8.1: Use the Hopfield and Hinton-Sejnowski formulas for storing the pattern $\mathbf{x}$ = (1 0 1).

Solution:

$$c = \begin{bmatrix} 1 & 0 & 1 \\ 0 & 0 & 0 \\ 1 & 0 & 1 \end{bmatrix}$$

HOPFIELD'S RULE

SIX PATTERNS TO STORE IN A 9-BIT C.A.M.

$$w_{ij} = \sum_{k=1}^{6} [(2x_i^k - 1)(2x_j^k - 1)]/6$$

$$= 4c_{ij} - 2c_{ii} - 2c_{jj} + 1$$

$$= 4c_{ij} - 1/3$$

$$= \begin{cases} 4/6 - 1/3 = 1/3 & \text{if } c_{ij} = 1/6 \\ 0 - 1/3 = -1/3 & \text{if } c_{ij} = 0. \end{cases}$$

HINTON-SEJNOWSKI FORMULA

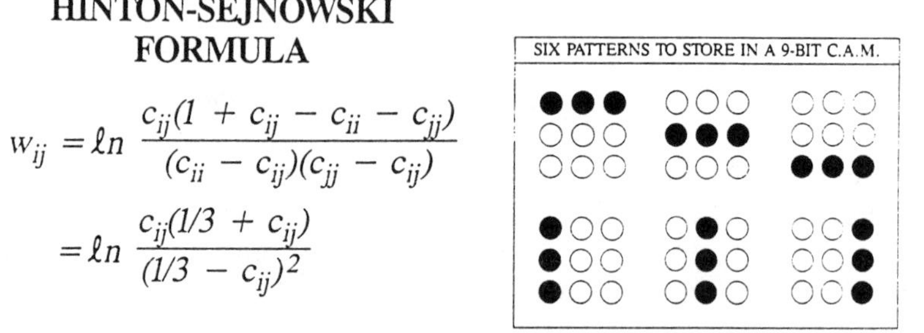

$$w_{ij} = \ell n \frac{c_{ij}(1 + c_{ij} - c_{ii} - c_{jj})}{(c_{ii} - c_{ij})(c_{jj} - c_{ij})}$$

$$= \ell n \frac{c_{ij}(1/3 + c_{ij})}{(1/3 - c_{ij})^2}$$

$$= \begin{cases} \ell n \dfrac{1/6 \cdot 3/6}{(1/6)^2} = \ell n\, 3 & \text{if } c_{ij} = 1/6 \\ \ell n\, (3\epsilon) \ll 0 & \text{if } c_{ij} = \epsilon \approx 0 \end{cases}$$

Figure 8.3 Alternative Storage prescriptions are applied to the same pattern set.

$$W = \begin{pmatrix} \ln\frac{1+\epsilon}{\epsilon} & \ln\frac{\epsilon}{1+\epsilon} & \ln\frac{1+\epsilon}{\epsilon} \\ \ln\frac{\epsilon}{1+\epsilon} & \ln\frac{1+\epsilon}{\epsilon} & \ln\frac{\epsilon}{1+\epsilon} \\ \ln\frac{1+\epsilon}{\epsilon} & \ln\frac{\epsilon}{1+\epsilon} & \ln\frac{1+\epsilon}{\epsilon} \end{pmatrix} .$$

Expanding the arguments of the logarithm in a MacLaurin series and keeping terms of order ϵ, $\frac{\epsilon}{1+\epsilon} \simeq \epsilon$:

$$W = \begin{pmatrix} -\ln\epsilon & \ln\epsilon & -\ln\epsilon \\ \ln\epsilon & -\ln\epsilon & \ln\epsilon \\ -\ln\epsilon & \ln\epsilon & -\ln\epsilon \end{pmatrix} = |\ln\epsilon|^3 \begin{pmatrix} 1 & -1 & 1 \\ -1 & 1 & -1 \\ 1 & -1 & 1 \end{pmatrix}$$

when $0 < \epsilon << 1$.

Problem 8.2: Use the Hopfield and Hinton-Sejnowski formulas to store a single pattern in a McCulloch-Pitts network.

Solution: The diagonal elements of the pattern correlation matrix are $c_{ii} = x_i$ when $\mathbf{x}$ is the single pattern. The off-diagonal elements are $c_{ij} = x_i x_j$. Thus $c_{ij} = 0$ or 1 for every i and j. Using equation (7),

$$W_{ij} = \begin{cases} +1 & \text{if } x_i = x_j \\ -1 & \text{if not} \end{cases} .$$

Using (6′):

$$W_{ij} = \begin{cases} |\ln(\varepsilon)| & \text{if } x_i = x_j \\ -|\ln(\varepsilon)| & \text{if not} \end{cases}$$

approximately, since $\epsilon/(1+\epsilon) \approx \epsilon$.

Figs. 8.4 (a—d) show some simple tests in which initial states and indicated inputs (biases) lead to stable states (with Glauber dynamics and McCulloch-Pitts neurons), completing the development started in fig. 8.1. It is evident that the storage of these six patterns in a nine-unit network produces some spurious stable states, especially when the outer product (Hopfield) rule is employed.

The Origins of the Storage Prescriptions

The origins of the storage prescriptions are as different as the Hopfield model and the Boltzmann machine. Chapter 1 already noted that the integration of an incremental, Hebbian learning formula (for the **W**'s) leads to the outer product rule (see p. 14). Hopfield (1982) pointed out that his model could be trained incrementally by taking **W** proportional to the summand in equation (1). The network might be given the ability to **forget** by writing:

$$c_{ij} \propto \sum_{t=1}^{\infty} x_i(t)\; x_j(t)\; \exp(-bt) \quad ,$$

for a "forgetting rate" $b>0$, t being the age of the "experience," **x**(t), imposed on the network by some environment. Then this (time- dependent) **c** is substituted in the storage prescription to obtain the current **W**.

Hinton and Sejnowski (1983) developed their formula for the weights of the collective computation network by considering a network of two McCulloch-Pitts units in the context of Bayesian inference using a Boltzmann machine.

<u>Problem</u> 8.3: A system of two sign neurons (Ising spins) has weight matrix **W**: $W_{12}=W_{21}=J$ and $W_{ii}=0$. The biases are y_1 and y_2. The system is in thermal equilibrium with a "heat bath" at temperature T. The joint density of the spins is given by the Boltzmann distribution:

$$P(s_1,s_2) = Z^{-1} \exp[(J\, s_1 s_2 + y_1 s_1 + y_2 s_2)/T]$$

where the partition function, Z, is a sum of four terms. Let:

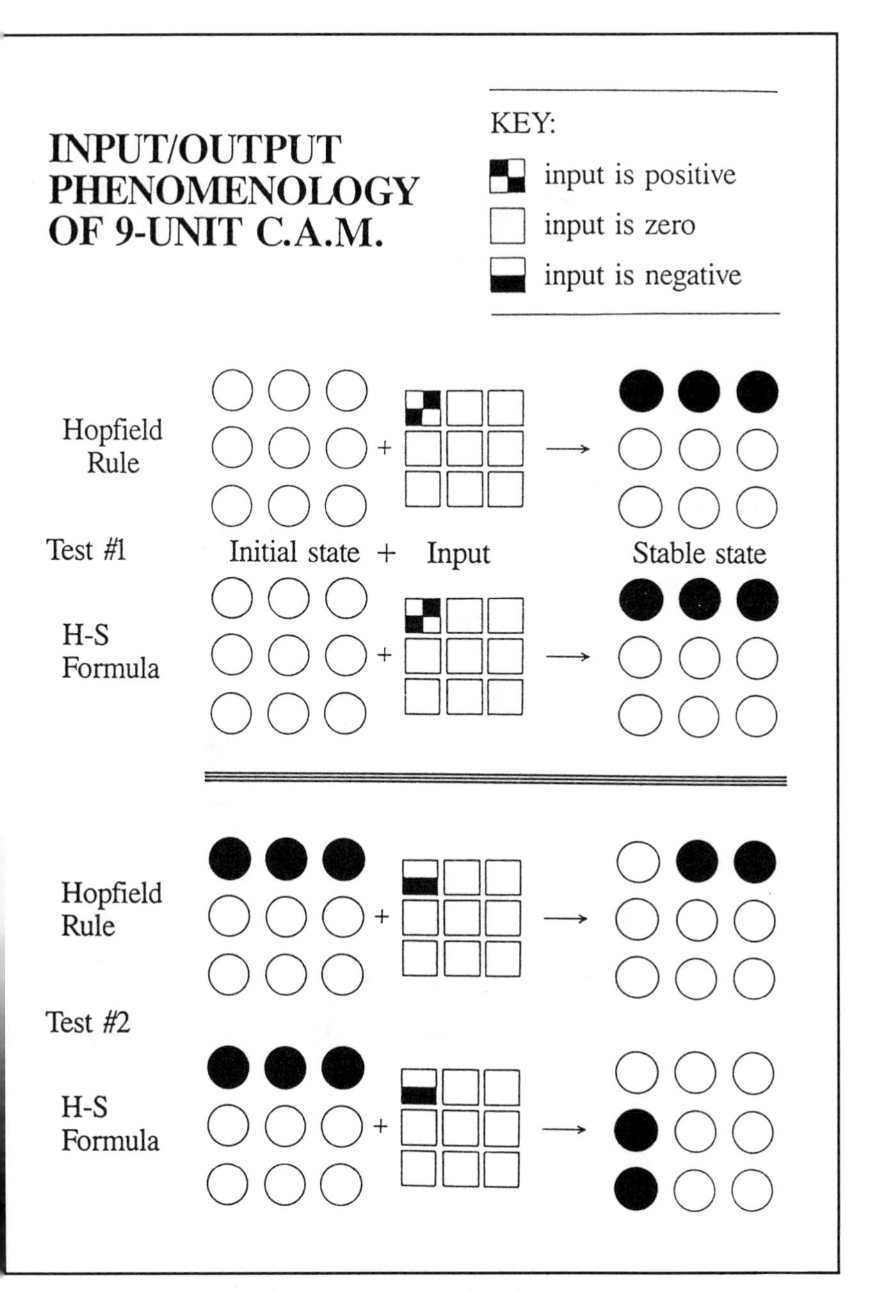

Figure 8.4a Input/Output phenomenology of a 9-unit C.A.M.

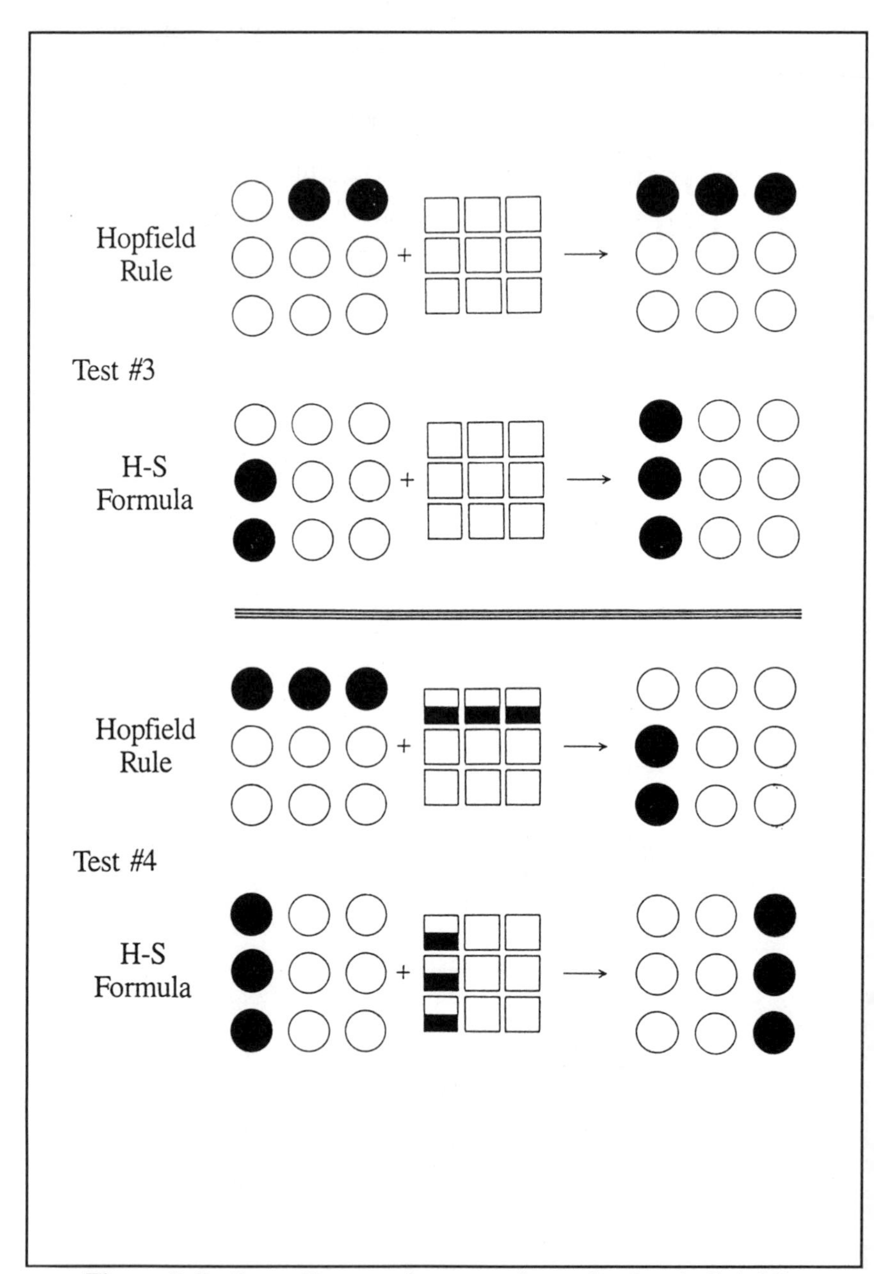

Figure 8.4b Input/Output phenomenology of a 9-unit C.A.M. (continued)

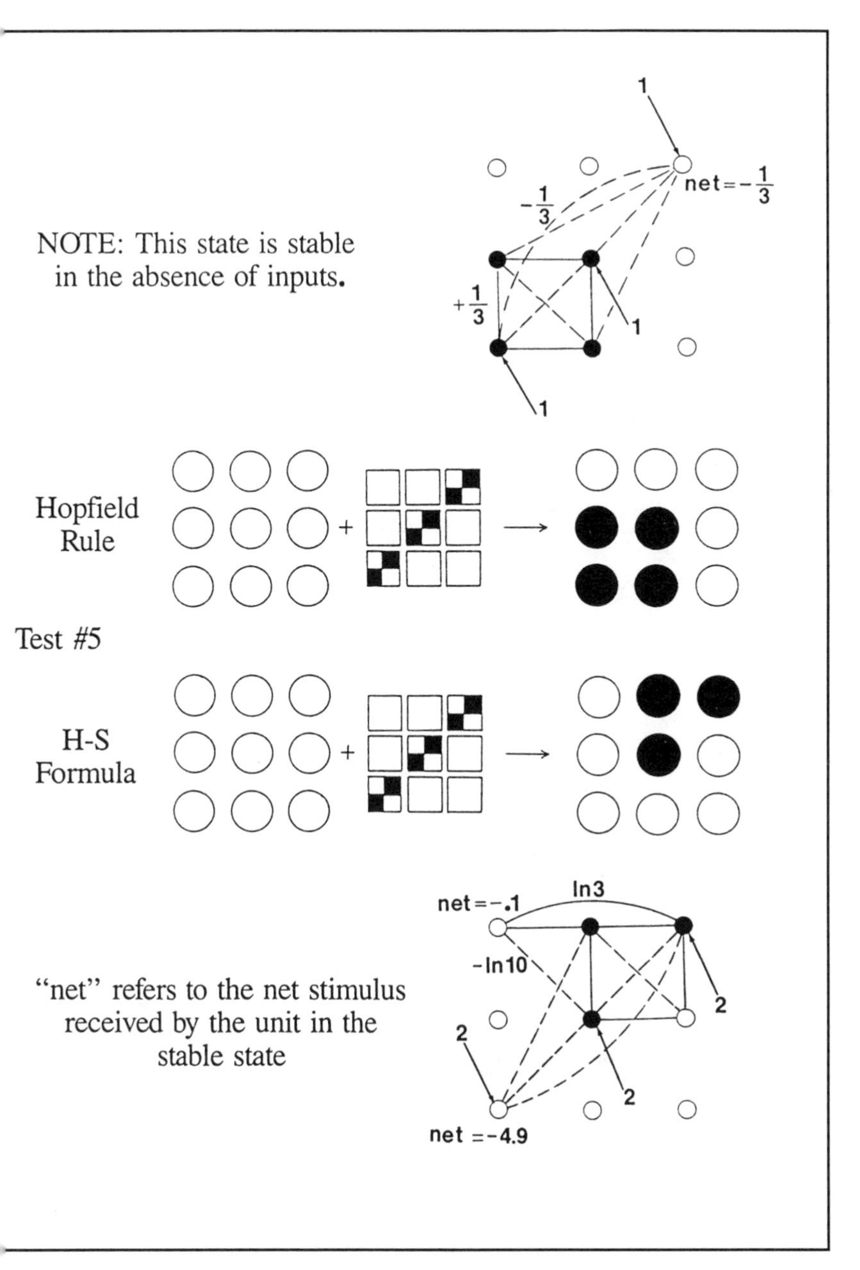

Figure 8.4c Input/Output phenomenology of a 9-unit C.A.M. (continued)

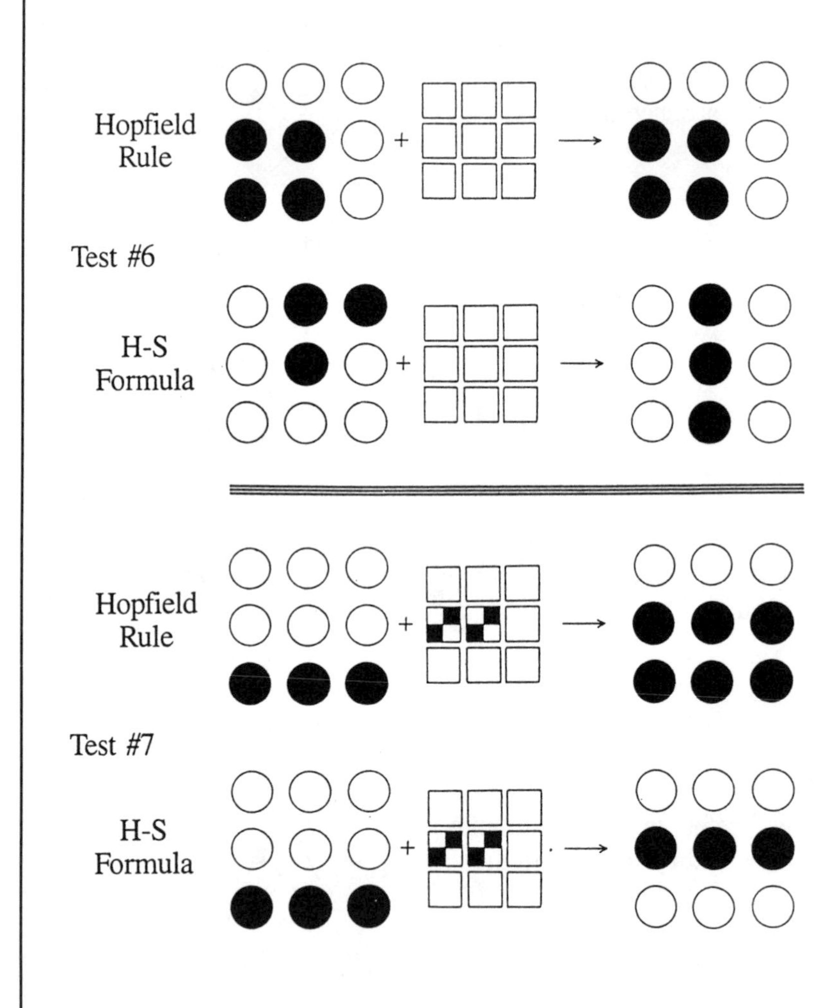

Figure 8.4d Input/Output phenomenology of a 9-unit C.A.M. (continued)

$$P = \begin{bmatrix} p_{00} & p_{01} \\ p_{10} & p_{11} \end{bmatrix} = \begin{bmatrix} p(-,-) & p(-,+) \\ p(+,-) & p)+,+) \end{bmatrix} = \begin{bmatrix} 1+c_{12}-c_{11}-c_{22} & c_{22}-c_{12} \\ c_{11}-c_{12} & c_{12} \end{bmatrix}$$

epresent this stationary joint density. If **c** is given, how must J, y_1, and y_2 be hosen in order to produce the given **c**?

Solution: Direct substitutions in the formula for the Boltzmann listribution give

$$P = \begin{bmatrix} \exp\,(J-y_1-y_2) & \exp\,(-J-y_1+y_2) \\ \exp\,(-J+y_1-y_2) & \exp\,(J+y_1+y_2) \end{bmatrix} /Z$$

t T = 1. These four equations readily yield the two identities:

$$2J\text{-}2\ln Z = \ln(p_{00}p_{11}) \quad \text{and} \quad \text{-}2J\text{-}2\ln Z = \ln(p_{01}p_{10}).$$

ubtracting the second from the first gives:

$$4J = \ln[p_{00}p_{11}/(p_{01}p_{10})].$$

The appropriate substitutions for the p's in terms of the c's gives:

$$J = (1/4)\ln\{[c_{12}(1+c_{12}\text{-}c_{11}\text{-}c_{22})]/[(c_{11}\text{-}c_{12})(c_{22}\text{-}c_{12})]\},$$

familiar result. The biases are:

$$y_i = (1/4)\ln\{[c_{i\,j}(c_{i\,i}\text{-}c_{i\,j})]/[(c_{j\,j}\text{-}c_{i\,j})(1+c_{i\,j}\text{-}c_{i\,i}\text{-}c_{j\,j})]\}\ j \neq i.$$

The lesson in this exercise is that the Hinton-Sejnowski formula couples ach **pair** of neurons with reciprocal connections the strength of which it adjusts o produce the same statistical properties that the pattern set imposes on the pair. This local, pairwise independent replication of the pattern statistics gives rise to ollective behavior across the Markov field constituted by the network.

The capacity of the neural network CAM is usually defined as the reatest number of patterns that can be stored and retrieved as stable states, erhaps subject to a fidelity criterion. If N is the number of neurons, let $M_{max}(N)$ be the greatest number of N-bit patterns that can be addressed. The apacity is:

$$C = \lim_{N\to\infty} M_{max}(N)/N;$$

and it is commonly assumed that this ("thermodynamic") limit is commensurat with computational experience with N on the order of a hundred.

Hopfield (1982), using the outer product rule, found that C wa typically 0.15, a result that has since been corroborated by several studies i which the fidelity criterion was more explicitly formalized (cf. Komlos an Paturi, 1988). In other words, when a network of 100 binary units is used t store more than fifteen randomly-generated 100-bit patterns, the network stat is more likely to converge to a fixed point that was **not** stored than to one of th desired patterns. The disappointingly low capacity of the Hopfield model the reflects the cluttering of the energy landscape with spurious (local) minima th creation of which is the inevitable by-product of the storage procedure. This ha been widely perceived as a serious flaw in the model and as cause to b pessimistic about the future of collective computation networks as associativ memory devices.

Yet this problem is akin to one we discovered in chapter 5 regardin neural optimization networks; and, had it been required that the network find th deepest energy valley at **zero temperature**, we would have abandoned th subject then and there. Perhaps we **did** abandon the Hopfield model — but onl to rediscover it as a special case of the Boltzmann machine. Regarding th ability of the Boltzmann machine to seek out the stored patterns in the midst o clutter, when the static capacity C is exceeded, virtually no authoritative wor seems to have yet been published.

Boltzmann Tic-Tac-Toe

The game of tic-tac-toe has long been used as an exercise by teacher of introductory courses on computing practice. Rumelhart et al. (1987, vol. 2 offer a neural network solution involving sixty-seven units that are (for the mos part) asymmetrically interconnected; and their description of the net may b lacking in quantitative completeness. Here we shall design an eleven-uni Boltzmann machine (one unit for each square and two "hidden" units — see fig 1.5) from which competitive moves can be elicited.

Fig. 8.5(a) shows the eight winning patterns on a tic-tac-toe board tha is augmented by two more squares one of which is dedicated to each diagona victory. The units of the basic 3x3 board are indexed in 5(b) in the manner o a magic square so that each row, each column, and each diagonal sum to fiftee (the extra squares ignored). (To the game theorist, the objective of tic-tac-to is to pick integers from the first nine, without replacement, so as to obtain a tri whose sum is 15.) Fig. 8.5(c) shows the pattern correlation matrix **c** (multiplie

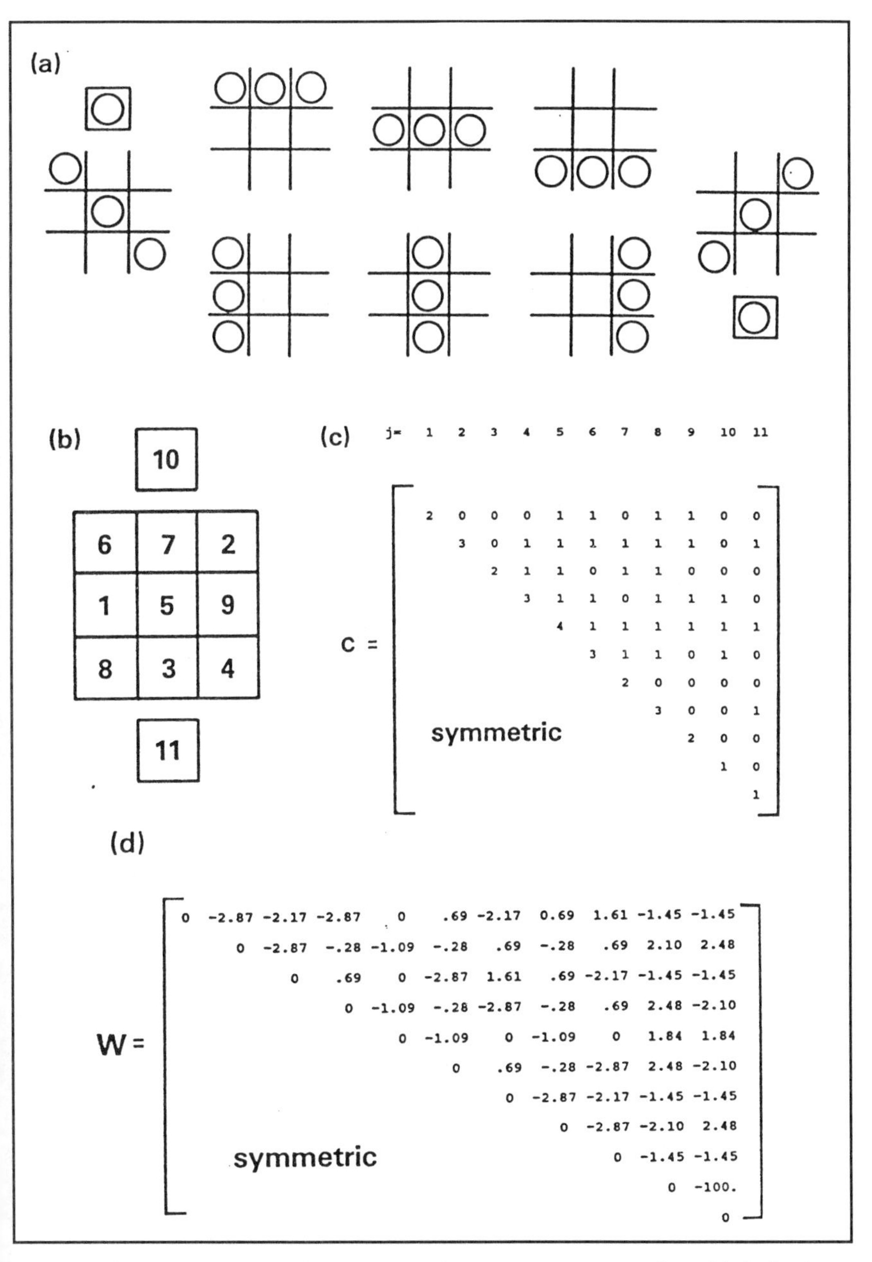

Figure 8.5 Boltzmann tic-tac-toe: (a) pattern enumeration,(b) indexing of units, (c) core relation matrix,and (d) connectivity matrix.

by the number of patterns) or, more precisely, the upper half of **c**, since **c** is symmetric. The Hinton-Sejnowski formula, equation (6'), is used to compute the weight matrix **W** from **c**. (Some "fixes" were subsequently made: $c_{10,10} = c_{11,11} = 1.2$ and $w_{10,11} = w_{11,10} = -100$.) In this problem it is necessary to use equation (2b) for the biases.

It remains to specify how an input represents a situation on the board and how the network is used to obtain moves. Define two interrogation phases, the X-plus and the O-plus, as follows: In the X-plus phase (ϕX+), the inputs are:

$$z_i = \begin{cases} +12 & \textit{if there is an "X" at position } i \\ 0 & \textit{if there is no marker at } i \\ -12 & \textit{if there is an "O" at position } i \end{cases}$$

In ϕO+ the signs are reversed. For each phase, iterate the familiar Glauber dynamical equation at a temperature $T > 0$ until some number K of random sweeps has been completed. After each sweep in ϕX+, survey the network state and increase the number n(i,X+) by one if and only if the ith McCulloch-Pitts unit is ON:

$$n(i,X+) \leftarrow n(i,X+) + x_i(11t), \quad t=1,\ldots,K.$$

This procedure leads to a 3x3 activity table:

$$\mathbf{n}(X+) = \begin{bmatrix} n(6,X+) & n(7,X+) & n(2,X+) \\ n(1,X+) & n(5,X+) & n(9,X+) \\ n(8,X+) & n(3,X+) & n(4,X+) \end{bmatrix}$$

which tells how many times each "visible" unit was ON in the course of K sweeps, as shown in fig. 8.6 (top). Following the same steps (but with signs reversed) an activity table **n**(O+) is obtained. The move is decided with reference to these tables. To place an "X, survey **n**(O+) and find the unoccupied position i^* for which the activity $n(i^*,O+)$ is greatest. To place an "O," survey the **n**(X+). The effect of these rules is to block the opponent's attempt to complete a winning pattern.

Fig. 8.6 (bottom) shows a solitaire game of Boltzmann tic-tac-toe, carried out with K=100 at temperature T=1.5, with the activity tables shown above and below the moves they direct. This game invokes a tie-breaking rule

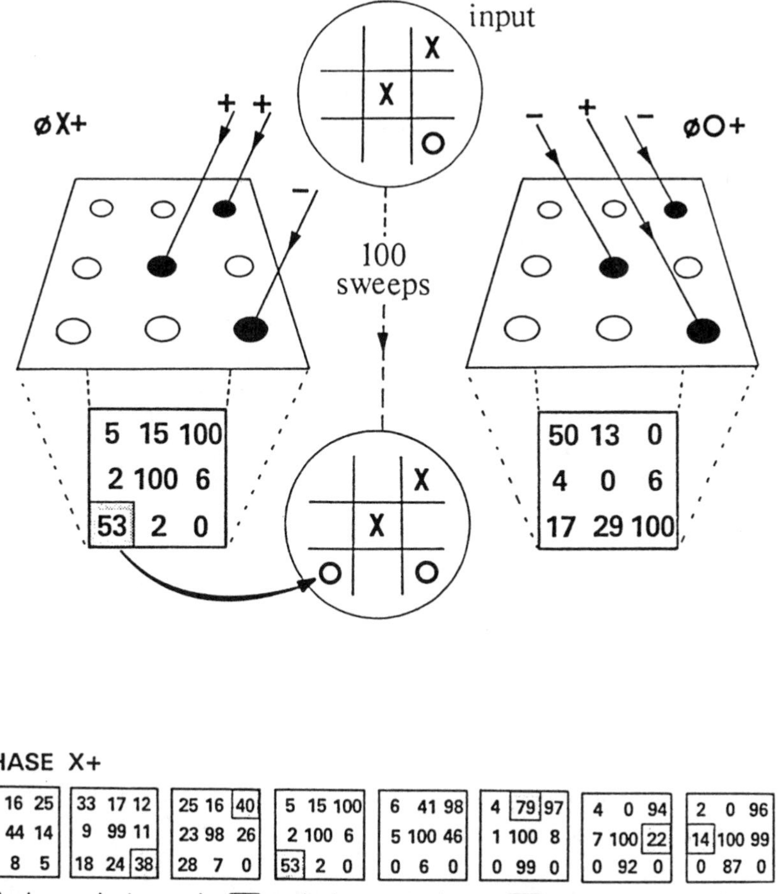

Figure 8.6 Boltzmann tic-tac toe: A sample game

at the third stage where the O+ table is ambivalent as to which corner should be taken by "X." The tie is broken with reference to activity table of the opposite phase. The rules are broken at the seventh stage where the controlling phase (O+) prescribes a move ("X" to position 7), which draws the game, overlooking a different move (to position 9), which gives "O" a final chance to make a fatal error. The alternative move is the one prescribed by the opposite phase; and the chance to lose is declined by "O" in the eighth and final stage.

Data Representation

Data representation in neural networks is typically inefficient in the sense that the number of binary units used to represent a data item is typically much greater than the number of bits that could represent it in some compressed form. The generous apportionment of neurons is the rule rather than the exception in PDP models. NETtalk, for instance, used twenty-nine input units per letter of text and twenty-six output units per phoneme when (obviously) eight bits would suffice in a minimal representation of either type.

Consider the "Jets and Sharks" problem in the first chapter of Rumelhart et al. (1987), recalled in part by fig. 8.7. Each gangster is characterized by a string of fourteen binary digits, each corresponding to a binary unit. The AFFILIATION of the individual is either 10 ("Jet") or 01 ("Shark"). MARITAL STATUS is represented by three units ("single," "married," or "divorced"). Obviously it takes only one bit to define AFFILIATION and two bits for MARITAL STATUS. Yet this proliferation of units makes possible the "strong lateral inhibition" that leads to a "winner- take all" competition among the units of the subnets that represent AFFILIATION MARITAL STATUS, etc.

Fig. 8.8 reduces the problem to fourteen units by dispensing with th "hidden units" and the name indicators. This reduced representation is furthe detailed by fig. 8.9a. If the problem is to be solved with a symmetric weigh matrix, strong lateral inhibition will be required within each of the five subnets Intuitively we can see that the unit representing the age bracket "30s" mus excite the unit indicating "Shark," since all the Sharks are in this age bracke Similarly, the unit for "Jet" must inhibit (and be inhibited by) the unit for "hig school" education, since each of the four named Jets either dropped out or wer to college.

These desiderata are realized by applying the Hinton-Sejnowski formul to the computation of the weights, as illustrated in fig. 8.9b. In particular, th

Name	Gang	Age	Edu.	M.S.	Occu.
Art	Jets	40s	J.H.	Sing.	Pusher
Al	Jets	30s	J.H.	Mar.	Burglar
Sam	Jets	20s	Col.	Sing.	Bookie
Clyde	Jets	40s	J.H.	Sing.	Bookie
Phil	Sharks	30s	Col.	Mar.	Pusher
Ike	Sharks	30s	J.H.	Sing.	Bookie
Nick	Sharks	30s	H.S.	Sing.	Pusher
Don	Sharks	30s	Col.	Mar.	Burglar

Network Diagram for Solving McClelland's Gangsters Problem, After Rumelhart et al. Note Visible Subnets are Linked to the Hidden (Central) Subnet Only.

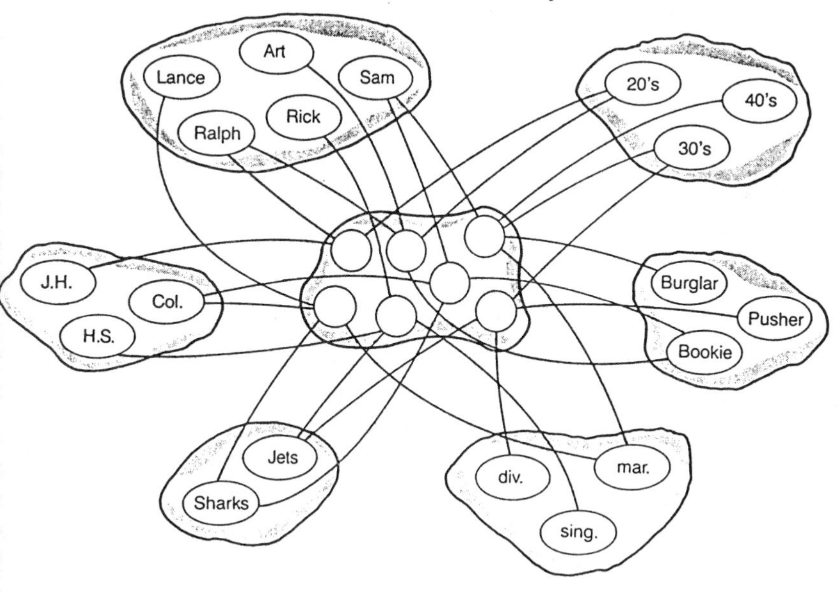

Figure 8.7 The Jets and the Sharks (from Rumelhart et al. (1986).

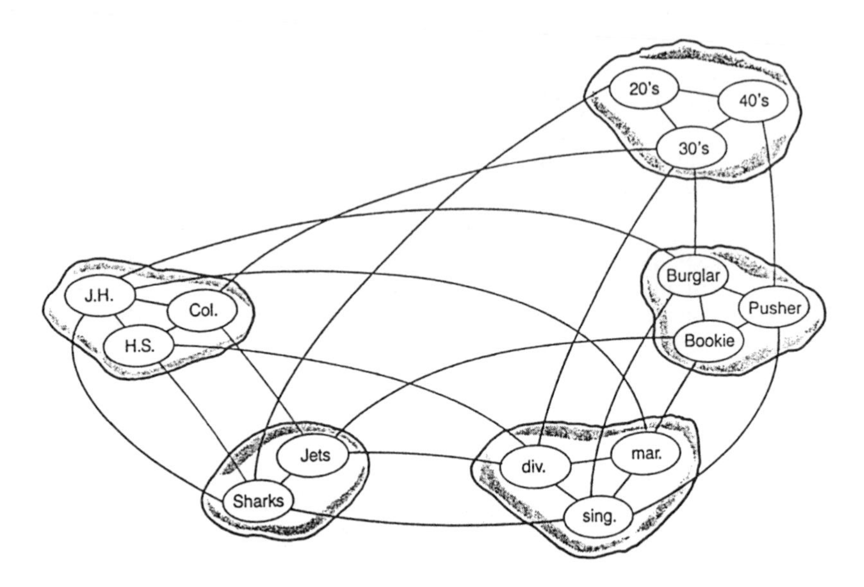

Figure 8.8 Network for solving gangsters problem reduced to fourteen units.

strength of the inhibitory links between units in the same subnet diverges as $\ln(\epsilon)$.

Fig. 8.10 shows the results of some tests using a rather crude BASIC program to go from initial states, input via keyboard, to final states, using the Glauber dynamics at zero temperature. The question marks in the input indicate unclamped units. In example 1, the input is a characterization of gangster "Art," complete except for AFFILIATION. The network quickly converges to a stable state in which the pattern is completed.

Strong Lateral Inhibition

Strong lateral inhibition, so important in satisfying the constraints of optimization problems in chapter 5, can also play an important role in content-addressable memories when the data items are suitably represented, as McClelland's gangster problem illustrates. This strong inhibition will arise naturally when the data representation allows it—and when the Hinton-Sejnowski formula is used for pattern storage.

Alternate Representation of Gangster Data

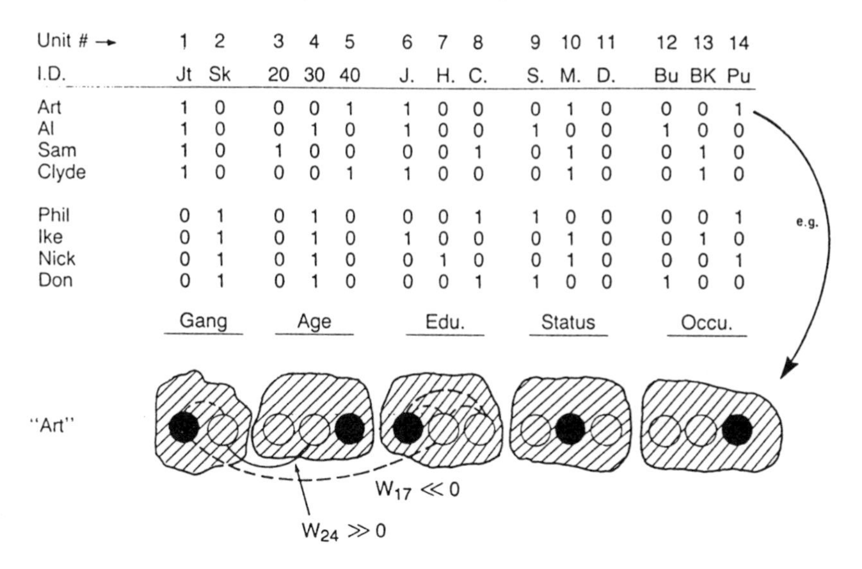

Unit # →	1	2	3	4	5	6	7	8	9	10	11	12	13	14
I.D.	Jt	Sk	20	30	40	J.	H.	C.	S.	M.	D.	Bu	BK	Pu
Art	1	0	0	0	1	1	0	0	0	1	0	0	0	1
Al	1	0	0	1	0	1	0	0	1	0	0	1	0	0
Sam	1	0	1	0	0	0	0	1	0	1	0	0	1	0
Clyde	1	0	0	0	1	1	0	0	0	1	0	0	1	0
Phil	0	1	0	1	0	0	0	1	1	0	0	0	0	1
Ike	0	1	0	1	0	1	0	0	0	1	0	0	1	0
Nick	0	1	0	1	0	0	1	0	0	1	0	0	0	1
Don	0	1	0	1	0	0	0	1	1	0	0	1	0	0
	Gang		Age			Edu.			Status			Occu.		

Storing the Gangster Data with the Hinton-Sejnowski Formula

Unit # →	1	2	3	4	5	6	7	8	9	10	11	12	13	14
I.D.	Jt	Sk	20	30	40	J.	H.	C.	S.	M.	D.	Bu	Bk	Pu
Art	1	0	0	0	1	1	0	0	0	1	0	0	0	1
Al	1	0	0	1	0	1	0	0	1	0	0	1	0	0
Sam	1	0	1	0	0	0	0	1	0	1	0	0	1	0
Clyde	1	0	0	0	1	1	0	0	0	1	0	0	1	0
Phil	0	1	0	1	0	0	0	1	1	0	0	0	0	1
Ike	0	1	0	1	0	1	0	0	0	1	0	0	1	0
Nick	0	1	0	1	0	0	1	0	0	1	0	0	0	1
Don	0	1	0	1	0	0	0	1	1	0	0	1	0	0
8	4	4	1	5	2	4	1	3	3	5	0	2	3	3

$C_{6,6} = 4/8$

$C_{6,10} = 3/8$

$C_{10,10} = 5/8$

For example, the symmetric weights linking units 6 and 10 have strength

$$W_{ij} = \log \frac{C_{ij}(1 + C_{ij} - C_{ii} - C_{jj})}{(C_{ii} - C_{ij})(C_{jj} - C_{ij})} = \log \frac{3\,(8 + 3 - 4 - 5)}{(4 - 3)(5 - 3)} = \log 3.$$

Figure 8.9 Data representation and pattern storage for The Sharks and Jets problem

Jets and Sharks: Example 1

Unit # →	1	2	3	4	5	6	7	8	9	10	11	12	13	14
I.D.	Jt	Sk	20	30	40	J.	H.	C.	S.	M.	D.	Bu	Bk	Pu
Art	1	0	0	0	1	1	0	0	0	1	0	0	0	1
Al	1	0	0	1	0	1	0	0	1	0	0	1	0	0
Sam	1	0	1	0	0	0	0	1	0	1	0	0	1	0
Clyde	1	0	0	0	1	1	0	0	0	1	0	0	1	0
Phil	0	1	0	1	0	0	0	1	1	0	0	0	0	1
Ike	0	1	0	1	0	1	0	0	0	1	0	0	1	0
Nick	0	1	0	1	0	0	1	0	0	1	0	0	0	1
Don	0	1	0	1	0	0	0	1	1	0	0	1	0	0
Input:	?	?	0	0	1	1	0	0	0	1	0	0	0	1
Output:	1	0	0	0	1	1	0	0	0	1	0	0	0	1

Interpretation: Gang (Art) = Jets

Example 2

Input:	?	?	0	1	0	?	?	?	?	?	?	?	?	?
Output:	0	1	0	1	0	0	1	0	0	1	0	0	0	1

Interpretation: If Age(Suspect) = 30s
Then Gang(Suspect) = Sharks [Nick]

Example 3

Input:	?	?	0	1	0	?	?	?	?	?	?	0	?	?
Output:	0	1	0	1	0	0	1	1	1	1	0	0	0	1

Interpretation: If Age(Suspect) = 30s
And Occupation(Suspect) = Not Burglar
Then Gang(Suspect) = Sharks [No Pos. I.D.]

Example 4

Input:	?	?	0	1	0	1	0	0	?	?	?	?	?	?
Output:	0	1	0	1	0	1	0	0	1	0	0	1	0	0

Interpretation: If Age(Suspect) = 30s
And Ed.(Suspect) = J.H.
Then Gang(Suspect) = Sharks [No Pos. I.D.]

Input:	?	?	?	?	?	?	?	?	?	?	?	?	?	?
Output/ Interpretation:	0	1	0	1	0	0	0	1	1	0	0	1	0	0

Pick a Suspect Given Initial State. Result is [Don].

Figure 8.10 Example of the network defined by the previous figure.

Problem 8.4: Some patterns are to be stored in a collective computation net of N units. Suppose the diagonal elements of **c** are all equal to 1/2. (This would be approximately the case when each pattern is generated by N independent Bernoulli trials with success probability 1/2, in the limit of large N.)

(a) Using the outer product rule, express $W_{ij} = w^{Hop}$ in terms of the corresponding element of **c**.

(b) Simplify the Hinton-Sejnowski formula to the extent possible, writing $W_{ij} = w^{H\text{-}S}$ in terms of correlation matrix components.

(c) Express $y = w^{H\text{-}S}$ as a function of $x = w^{Hop}$.

Solution:

(a) $w^{Hop} = 4c_{ij} - 2c_{ii} - 2c_{jj} + 1 = 4c_{ij} - 1$.

(b) $w^{H\text{-}S} = -(1/2) \ln[1/(2c_{ij}) - 1]$ after multiplying W_{ij} by 1/4 as suggested by problem 8.3.

(c) The result of the first part shows that $c_{ij} = (1/4)(w^{Hop} + 1) = (x+1)/4$. Substituting this in the second expression gives $y(x) = -(1/2) \ln[2/(x+1) - 1]$.

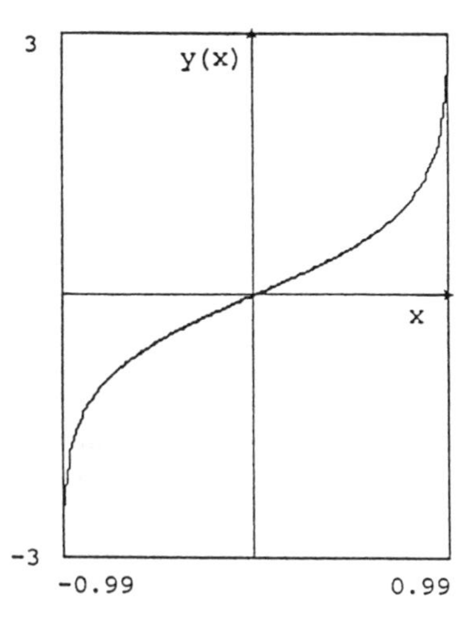

Problem 8.5: Some patterns are to be stored in a network of N units. Half of the patterns have $x_1 = 1$ (unit #1 ON) and the other half have $x_2 = 1$. There is no pattern for which $x_1 = x_2$. For every pattern, $x_3, \ldots, x_N$ are all generated by a random coder that sets $x = 0$ or 1 with the same probability (1/2). Thus the network should serve as a two-category classifier when the first two units are treated as output units and the last N-2 are input units.

(a) Evaluate W_{12} using the Hopfield and Hinton-Sejnowski formulas. In the latter case, let $\epsilon = 1/N$.

(b) Evaluate W_{13} using the same formulas. Assume $c_{33} = 1/2$.

(c) Evaluate W_{12}/W_{13} in the limit of large N for each case.

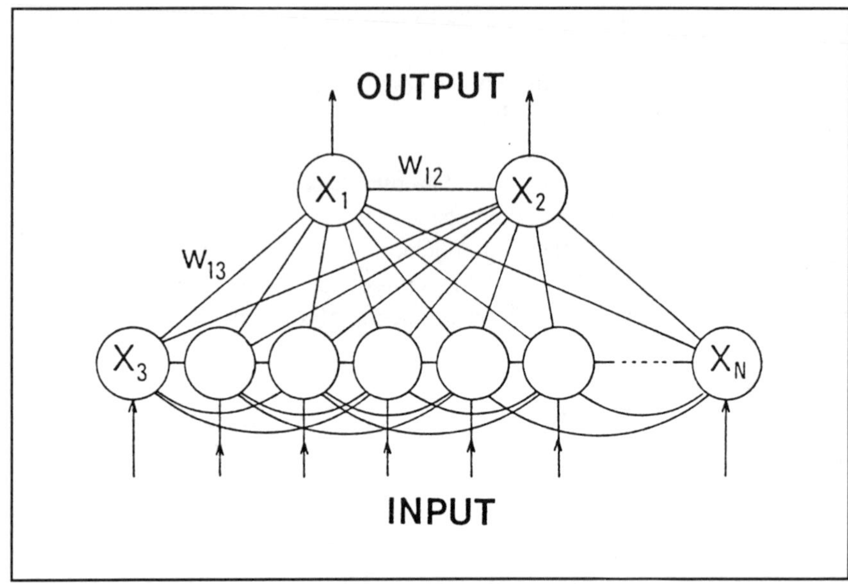

Binary Classifier

Solution:

(a) For W_{12} obtain -1 with the outer product rule and - ½ ln(2N) with the Hinton-Sejnowski formula.

(b) With reference to the preceding problem, W_{13} takes the values $4c_{13}$ - 1 and -(1/2)ln[1/($2c_{13}$) - 1].

(c) If c_{13} is a fixed, positive quantity between 0 and 1/2, then the second expression above is finite. In the limit of large N, however, the inhibition between units 1 and 2 grows infinitely strong when the Hinton-Sejnowski formula is used: $W_{13}/W_{12} \rightarrow 0$ as $N \rightarrow \infty$. With Hopfield's prescription, on the other hand, the same ratio remains finite. (See the preceding problem.)

Additional Problems

8.6 Let the diagonal components c_{ii} and c_{jj} of the pattern correlation matrix both equal 1/2. Simplify the expressions for the weight W_{ij} in terms of c_{ij} using

(a) the outer product rule and

(b) the Hinton-Sejnowski formula.

(c) Calculate the correlation coefficient ρ_{ij} and compare to the result of part (a).

8.7 Derive a storage prescription by equating the covariance $E(x_i x_j) - (Ex_i)(Ex_j)$, with reference to the pattern set, to the equilibrium covariance of the pair (x_i, x_j) at a temperature $T=1$.

8.8 Let $(x_1,\ldots,x_N)$ be the state of a Boltzmann machine composed of sign neurons at temperature $T > 0$. Let $p_{ij} = \Pr(x_i=+1,x_j=+1)$. Prove the following two propositions:

(a) When $N \geq 3$, $p_{ij} = (1/4)(1 + T\,\partial \ln(Z)/\partial W_{ij})$, where Z is the partition function.

(b) When $N = 2$, $p_{12} = (1/4)[1 + \tanh(W_{12}/T)]$.

8.9 The objective is to store <u>four</u> five-bit patterns:

pattern #		x_1	x_2	x_3	x_4	x_5
	1)	0	0	0	1	1
	2)	1	0	0	1	1
	3)	0	1	0	1	1
	4)	1	1	1	0	0

in a network of McCulloch-Pitts neurons.

(a) Show the upper-right triangle and the diagonal components of the pattern correlation matrix. Recall that c_{ij} is the fraction of patterns for which units i and j are both ON.

(b) Use Hopfield's storage prescription to fill in the weights on the symmetric links of the network diagram:

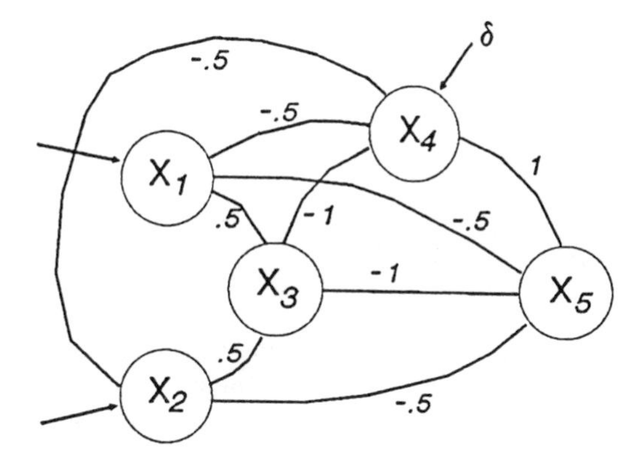

(c) Let the state be 00000 initially. Now apply a very small stimulus $\delta > 0$ to unit #4. What stable state is reached?

(d) Now apply a very large stimulus to unit #1 so that it is clamped ON irrespective of the states of the other units. Do any other units now turn ON?

(e) Clamp unit #2 to the ON state (as well as #1). What stable state is reached?

(f) Treating units #1 and #2 as inputs, and #5 as the output, what elementary logic function is performed by the network? What if #4 is the output?

(g) Returning to the pattern correlation matrix, use the Hinton-Sejnowski formula to compute the weights. (Use equation (6') on page 162.) Let g = $-\ln(\epsilon)$. Divide all the weights by g and compare the result to that of part (b) in the limit $\epsilon \rightarrow 0$.

Solution:

(a)

$$\boldsymbol{c} = \begin{pmatrix} 2 & 1 & 1 & 1 & 1 \\ & 2 & 1 & 1 & 1 \\ & & 1 & 0 & 1 \\ & & & 3 & 3 \\ & & & & 3 \end{pmatrix}$$

(b)

$$\boldsymbol{W} = \begin{pmatrix} 0 & 0 & .5 & -.5 & -.5 \\ & 0 & .5 & -.5 & -.5 \\ & & 0 & -1 & -1 \\ & & & 0 & 1 \\ & & & & 0 \end{pmatrix}$$

(c) State 00011 is stable.
(d) No change.
(f) NAND and AND.
(g) In the limit, the result is the same as (b).

8.10 Let seven McCulloch-Pitts neurons be in one-to-one correspondence with the segments of a numeric display as shown in the sketch. The numerals 0 through 9 are thus represented by the following seven-bit codes:

(a) List the diagonal elements of the pattern correlation matrix, **c**.
(b) Compute W_{46} using formula (6').

unit index

		1 2 3 4 5 6 7
"0"	=	1 1 1 1 1 1 0
"1"	=	0 0 1 1 0 0 0
"2"	=	0 1 1 0 1 1 1
"3"	=	0 1 1 1 1 0 1
		... etc. ...
"9"	=	1 1 1 1 1 0 1

c) Since there are ten patterns but only seven units, it will not be surprising to find that the patterns do not correspond to stable states and vice versa. Can this situation be improved by dedicating two units to each segment? Three units?

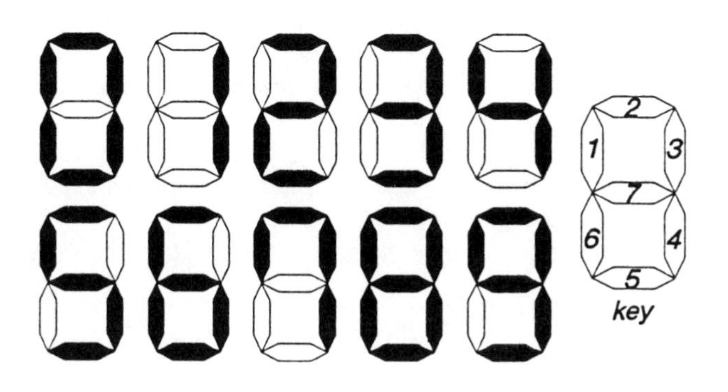

Ten patterns are represented by seven units as indexed in the key above. Solid units are ON.

Solutions:

(a) $[c_{11}, \ldots, c_{77}] = [.7, .8, .8, .9, .7, .4, .7]$.

(b) Note $c_{46} = 0.3$. Thus $W_{46} = \ln[(.3)(1+.3+\epsilon-.9-.4)] - \ln[(.9-.3)(.4-.3)] = \ln[.3\epsilon/.06] = \ln(5\epsilon)$ for $0 < \epsilon << 1$.

(c) Adding redundancy in this manner does nothing to improve the network's ability to recall the desired patterns. Numerical experimentation readily shows that the energy landscape of the original (seven-unit) net is dominated by a broad basin of attraction about the state 1100111, which corresponds to "E" (as in error). Dedicating more units to each segment just makes the "E" grow bigger as suggested by this sketch.

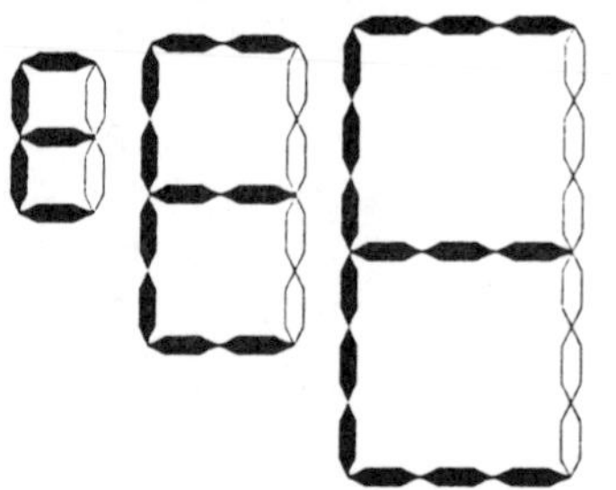

8.11 Let there be some N-bit patterns each of which has exactly K nonzero bits. Suppose that N/K = M is an integer and that M is the number of patterns. Mapping the pattern bits into binary units, assume that the patterns do not overlap and compute the weights and biases using the Hinton-Sejnowski formula. (Since the patterns do not overlap,

$$c_{ij} = \begin{cases} 1/M & \textit{if } i \textit{ and } j \textit{ both turn on the same pattern} \\ 0 & \textit{otherwise} \end{cases}$$

Are all the patterns stable? What is the minimum capacity of the network as far as K-bit, nonoverlapping patterns are concerned?

Solution: The diagonal elements of **p** are identically 1/M. Hence the weights are:

$$W_{ij} = \begin{cases} -2\ln(\varepsilon) + \ln(M) & \textit{if } c_{ij} = 1/M \\ \ln(\varepsilon) + \ln(M^2-2M) & \textit{if } c_{ij} = 0 \end{cases}$$

$0 < \epsilon << 1$. The bias is -ln(M - 1) for every unit. Setting $\epsilon << 1/M^2$ makes the patterns stable. Thus the capacity is at least M/N = 1/K where K = 2, 3, 4, ...

9. Emergent Properties of Neural Network Associative Memory

In a short course titled "Neural Networks for Artificial Intelligence (AI)," Hinton attributed a **structure** to neural network associative memory. With reference to fig. 9.1, this structure is "a set of subfields with emergent properties". Shall we accuse Hinton of trying to capture in a few words what we have spent the last eight chapters describing? In particular, the meaning of "emergent properties" triggers our curiosity. The figure depicts a network of neuronlike units partitioned into four subnets each of which can represent a "piece of the puzzle". The reciprocal connections of each pair of units in the

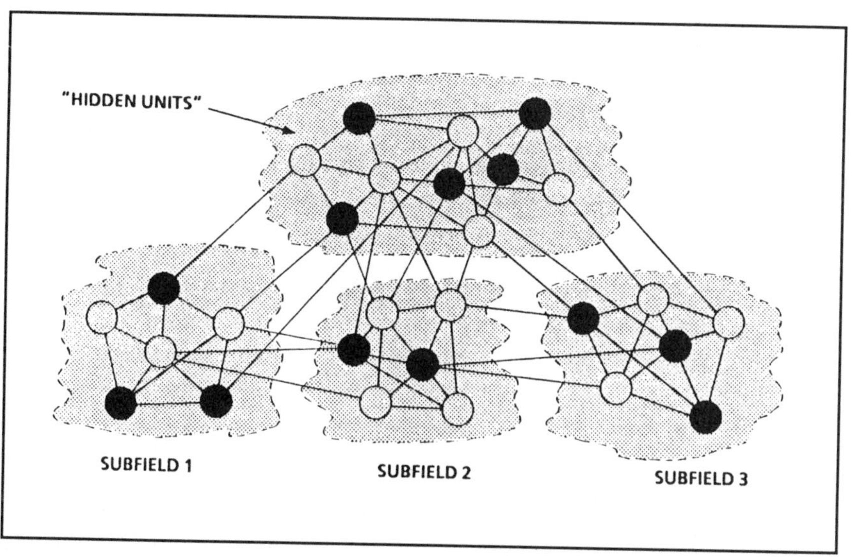

Figure 9.1 Structure is a set of subfields with emergent properties. Each subfield has stable patterns that represent possible "solutions." The pattern in one subfield influences all the others.

network were created by storing complete patterns (like solutions to the puzzle). Each subnet interacts with every other to minimize global energy. If the units of one or more subnets are clamped, the process of energy minimization leads to global pattern formation in which the stable states of the whole network are like solutions to the puzzle. Given partial information in the form of the clamps, the network completes the pattern. Does this exhaust the meaning of "emergent properties"?

Since the European Enlightenment, philosophers have advanced the reductionist view that human intelligence is essentially the manifestation of a vast and efficient associative memory device called the brain. To be sure, "associative memory device" was not in the vocabulary of David Hume when he pointed out that so much of the intellectual gift that humans enjoy can be rationalized in terms of the ability to associate objects in our environment —understood as the totality of our experience — through the medium of cognition. Serious attempts to recap the history of philosophy being put aside, we might simply trace the modern development of Hume's skeptical tradition to Ogden and Richards (1923), who wrote:

> The most celebrated of all caterpillars was striped yellow and black and was seized by one of [Prof. Lloyd Morgan's] chickens. Being offensive in taste to the chicken he was rejected. Thenceforth the chicken refrained from seizing similar caterpillars. Why? Because the sight of such a caterpillar, a part that is of the whole sight-seize-taste context of the original experience, now excites the chicken in a way sufficiently like the whole context did, for the seizing at least not to occur, whether the tasting (in images) does or not.
>
> This simple case is *typical* of all interpretation, the peculiarity of interpretation being that when a context has affected us in the past the recurrence of merely a part of the context will cause us to react in the way in which we reacted before.

Ogden and Richards attempted to reduce *The Meaning of Meaning* to the cognitive act of associating mental objects with sensory stimuli called words. In this view, the meaning of the term emergent properties could consist in making the association to: "Neural Networks and Physical Systems with *Emergent* Collective Computational Abilities" (Hopfield 1982). While the central nervous machinery required to make such an association is considerably more advanced than the chicken's, it might not be fundamentally different when reduced to its essential building blocks (and higher level structural regularities?).

The mainstream AI school of Minsky, in the 1960s and 1970s, followed the philosophical path illuminated by Kant, who rejected Hume's empiricism on the grounds that *a priori knowledge* is required to make use of environmental stimuli. Moreover, humans learn to develop *logical rules* for organizing perception and experience. These "synthetic a priori" truths made the inadequacy of the associative memory analogy self-evident (to perhaps express Kant's view in modern jargon).

Whether cheap, reliable, high-speed associative memory is indeed the key to developing "intelligent systems" cannot be posed as a scientific question. It is, however, a major issue in technology policy that will be addressed in at least one decade prior to the advent of machine intelligence — signified, according to Minsky, by the use of language commensurate with a six-year-old child's and estimated to lie perhaps 300 years in the future.

The transient eclipse of the neuromorphic approach to AI, pioneered (in the 1950s) by Farley (1960), Rochester (1956), Rosenblatt (1960), and others, coincided with advances in the fabrication of large-scale integrated circuits, leading to the microprocessor (1974) and the personal computer, which made a market for neurocomputer entrepreneurs (after symbolic processing lost its Kantian mandate and casual users realized that "inference engines" had to be driven by serious intellectual labor). Carr and Mize (1972) contrasted conventional computer memories and CAMs in this way:

The ROM and RAM are coordinate addressable memories, and data cannot be read or written until a particular, unique cell location is specified. In applications such as a search for a match to particular data items, all cells of the memory must be interrogated. There exists another class of memories which provides the necessary circuits for matching to a content specification instead of an address. This type of memory is termed content-addressable memory or CAM. These memories are also called associative memories. With CAM, the content desired for matching is address instead of location. We specify [the] data desired by category, and not by specific coordinate address.

The CAM becomes effective when we must search a mass of data in a parallel fashion. The ability to search out or interrogate stored data on the basis of content can be a powerful asset in many applications. Another facet sometimes overlooked is that the CAM may be the basis for a truly general associative data processor. The simple matching process can be iterated to perform complex operations of obtaining best match, next best match, less than, greater than, etc., following a single memory interrogation. The computer does not need an address code to keep track of words stored in the associative memory. Instead the computer specifies the *content* of the words its needs, and out come the words, wherever they are stored.

The cost-per-bit of associative memories is, of course, much greater than that of conventional memories. In the past, this increased cost has made it prohibitively expensive to utilize core memory in associative designs to any extent. At the present time, very little is known concerning the programming needed for this type of computer hardware. As simple systems are introduced into the marketplace and computer scientists learn more concerning the manipulation of these systems, we can expect more useful applications.

The technological question, then, is whether collective computation networks are a viable basis for associative processing. If they are, and if they can be produced at a competitive price, then the next generations of computer users will decide how much power resides in the ability to associate data items in ways restricted more by human cleverness than by device limitations.

As Ogden and Richards must have pointed out, part of our existential plight consists in the compulsion to put **names** to the things that pain us. Systems engineers have names that they append to functional requirements that they would like to satisfy but can't for lack of hardware and software tools. "Natural language processing", "machine vision", and "data fusion" are common examples. Many such names can be subsumed under the heading of "artificial intelligence". The view has been widespread since 1988, that neural network technology will have to quickly prove its ability to solve difficult AI problems or lose the policy debate and much of the largesse that goes with public initiatives and crash programs. If there are twelve unsolved problems under the aegis of AI (corresponding to the labors of Hercules), collective computation networks might be expected to solve one of them, thus vindicating Hopfield's view of biophysics as a new source of theory to drive the progress of advanced technology. The fact that most neural modelers coming from the biological side of the university find fault with the brain theory that Hopfield tried to develop with his model will be irrelevant to the technologist. The mathematics of collective computation is sufficiently well defined to leave few questions that cannot be tested scientifically in digital simulation models.

Naturally we look to these models to shed light on the future uses of ully analog collective computation networks. Baran (1989a, 1989b) claimed to ave performed "data fusion" (in the sense of multisensor integration) using a eural network associative memory based on the stochastic model to which this ext has been so devoted. The problem was one of parameter estimation on a

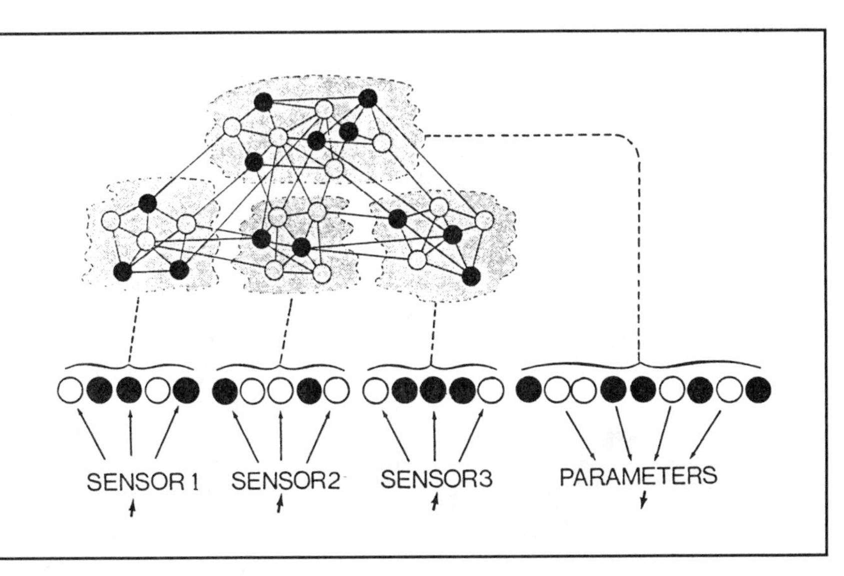

Figure 9.2 Interacting subnets associate sensory inputs with a parameter vector.

set of simultaneous waveforms originating in different sensors. Instead of decomposing the waveforms into sums of orthogonal (basis) functions, or training a backpropagation network to recognize spectral characteristics, Baran's model simply stored the waveforms in a CAM, together with the parameters defining them, using the Hinton-Sejnowski formula. Fig. 9.2 shows the same four subnets as fig. 9.1; but now they are given some functional significance. In particular, the first three subnets represent the reports of three sensors. The fourth is a coded representation of the parameters that generate the waveforms through parametric models. Each point in parameter space gives rise to a trio of waveforms. (Eg., each choice of the parameter σ in chapter 6, fig. 6.4 defines three distribution functions.) Thus each point of the lattice $\{\underline{\Theta}^m: m=1,2,\ldots,M\}$ in parameter space produces a set of patterns of the form $(s_1^m, s_2^m, s_3^m, \underline{\Theta}^m)$, where $s_k^m = (s_k(0;\underline{\Theta}^m),\ldots,s_k(L;\underline{\Theta}^m))$ is the sampled data representation of the waveform out of the kth sensor, at discrete times 0,1,...,L and $\underline{\Theta}^m$ describes the physical setting. Fig. 9.3 shows the subnets as grids upon which the pieces of the pattern:

$$\mathbf{x}^m \Leftrightarrow (\mathbf{s}_1{}^m, \mathbf{s}_2{}^m, \mathbf{s}_3{}^m, \underline{\Theta}^m)$$

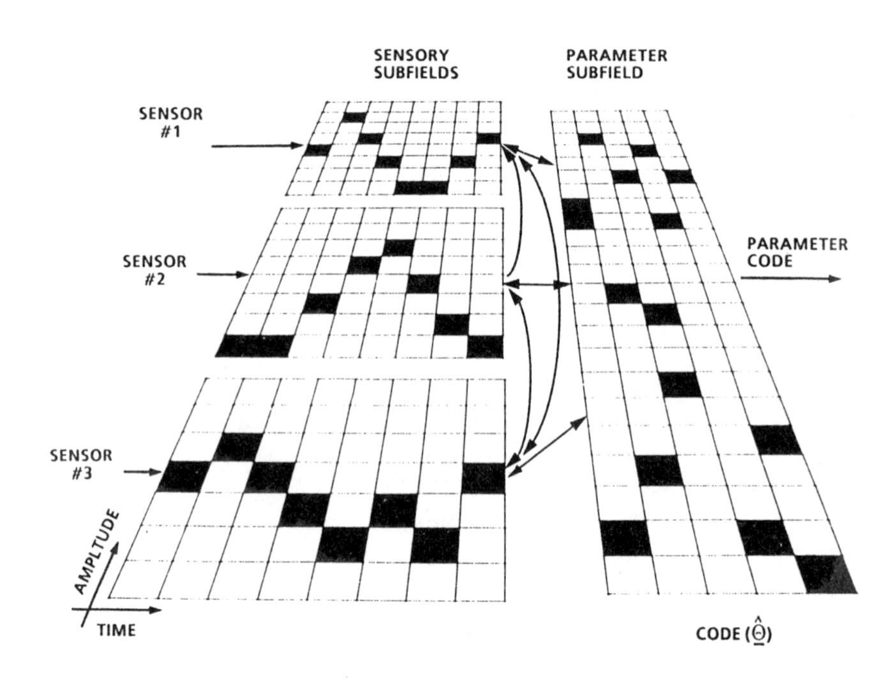

Figure 9.3 Data representation in binary subfields corresponding to subnets of a neural network associative memory.

are impressed, the filled boxes being McCulloch-Pitts neurons that are turned ON. After storing all M such patterns, clamping the units in the sensory subnets as shown in the figure should excite the parameter subnet *as shown* — if the CAM functions as a pattern completion device. In this way the neural network associates parameters with waveforms.

The fact that the number of patterns was not much smaller than the number of units caused spurious energy minima to proliferate. By raising and lowering the intensity of noise added to the input of each unit in the digital

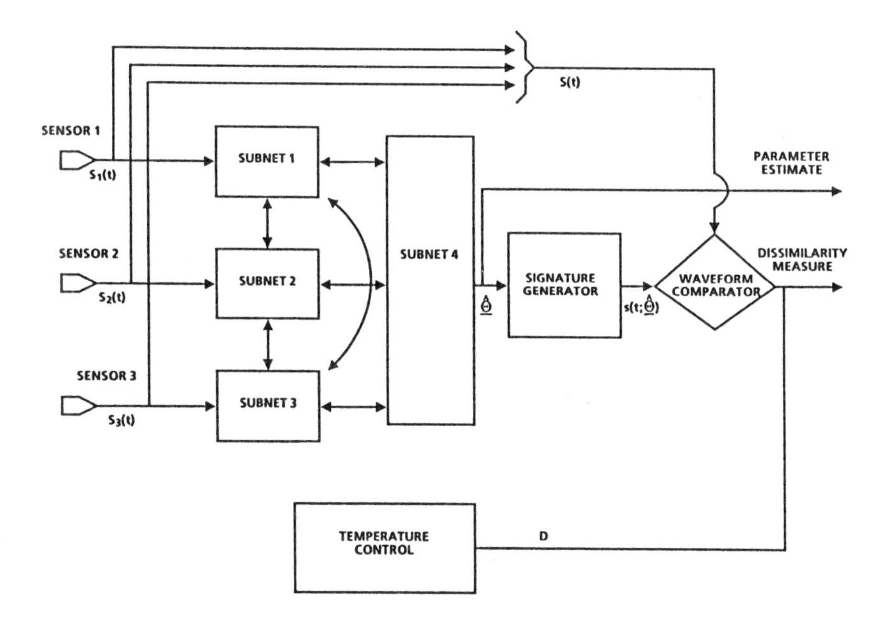

Figure 9.4 Block diagram of neural network associative memory model.

simulation model — thus controlling the temperature of the Boltzmann machine — Baran was able to find good (although suboptimal) solutions to the best match problem using the scheme of fig. 9.4. The neural network associative memory served as a match limiter, producing parameter estimates, and thereby nominating candidate waveforms for comparison to the sensor reports using an appropriate dissimilarity measure. When the parameter estimate is poor, the dissimilarity is high; and the temperature is raised in proportion to (the log of the) dissimilarity before cooling. When cooling leads to a good parameter estimate, the next heating cycle is more moderate. Baran claimed that the effectiveness of this approach was limited mainly by the speed and (RAM) storage capacity of the general purpose computer that hosted the digital simulation.

Concurrently Asynchronous Processing

The number of interconnects in the collective computation network is roughly the square of the number of units. Imagine a digital retina of N pixels that are one-to-one with the neurons. Clearly the most coarsely quantized image quickly fills the RAM of the computer with synaptic weights. Following the precedents noted in the preceding chapter, data compression as a preprocess to data representation is antithetical to the PDP outlook. Thus the network is slow and cumbersome to simulate with even fast computers of the present generation. Witness Crisanti and Sompolinsky (1988), who found it necessary to terminate the Glauber dynamical simulation of asymmetric nets by their Cray X-MP after only fifteen thousand sweeps — far fewer than the 2^N sweeps needed to search the state space of the network with $N > 100$.

The rise of concurrent processing and the popularity of local memory multiprocessors (like the Inmos Transputer) opens new possibilities for digital simulation models of collective computation networks. It has been reported that concurrently asynchronous processing in Transputer-based systems can expedite the zero temperature convergence of Hopfield nets by a factor proportional to the number of Transputers. Fig. 9.5 shows two ways of partitioning the fully interconnected network into subnets of roughly equal size. The subnets are assigned to processors that concurrently iterate the dynamical equation, selecting units at random from their own subnets, and treating the outside units as if they were clamped. The processors periodically stop iterating to share information, updating the full network state vector in their local memories.

When the N units are partitioned into B subnets of roughly equal size, there are still N^2 interconnects to consider; but B concurrent processors can each use their local memories to store the fewer than $(N/B)^2$ weights that are needed to iterate the dynamical equation within their respective subnets. Thus the 10^6 or so units on the retina of fig. 9.6 might give rise to only 10^4 interconnects per processor if there are 10^4 of these (pixel neighborhood) processors. Ten thousand concurrent processors might (even today) consume less than the 150 kilowatts used by the thirty ton ENIAC in von Neumann's time. The burden — and benefit — of full connectivity would not be realized until it is time for these 10^4 processors to shake hands and exchange new substates for old. The Markov field properties of the 10^6-unit collective computation network (in the pure sense) might not be too severely distorted by the partition so long as the global state is refreshed in the local memory of each processor at intervals less than the mean number of sweeps to (zero temperature) convergence (chapter 3). The more frequently this information is propagated across the whole retina, the

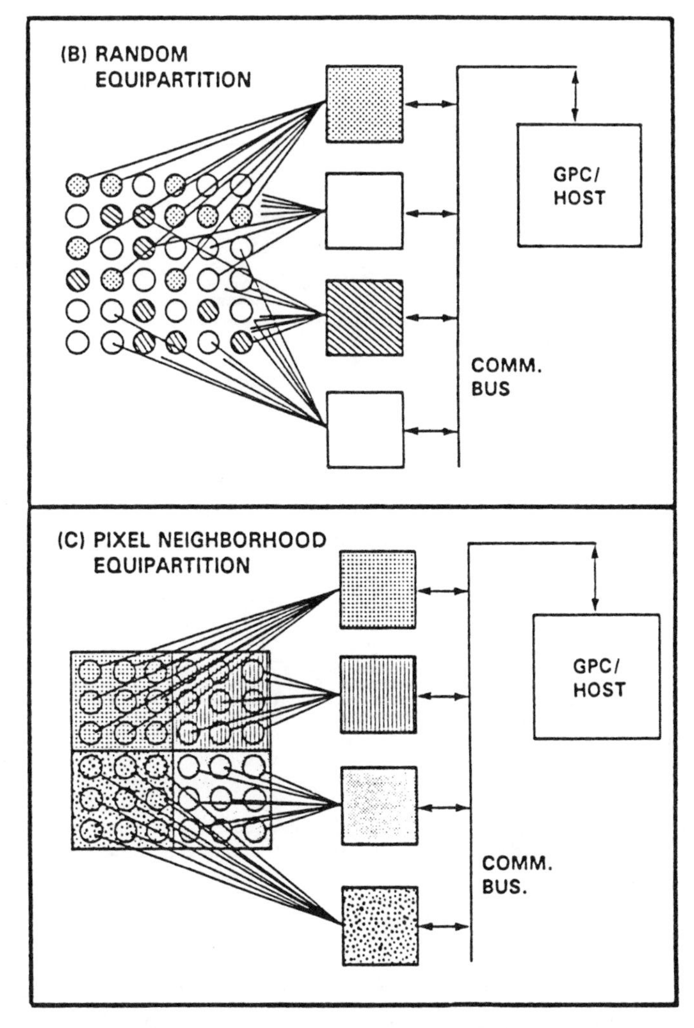

Figure 9.5 Two ways of partitioning the units for concurrently asynchronous processing.

longer it will take for the whole network to converge — assuming the total time is dominated by the communication phase.

Simpler examples of concurrently asynchronous network simulation can be used for illustration. Fig. 7.9 depicted, in effect, two Boltzmann machines of identical connectivity but different temperature. The warmer one drove the cooler one through the synapses made by each unit in the one net upon its

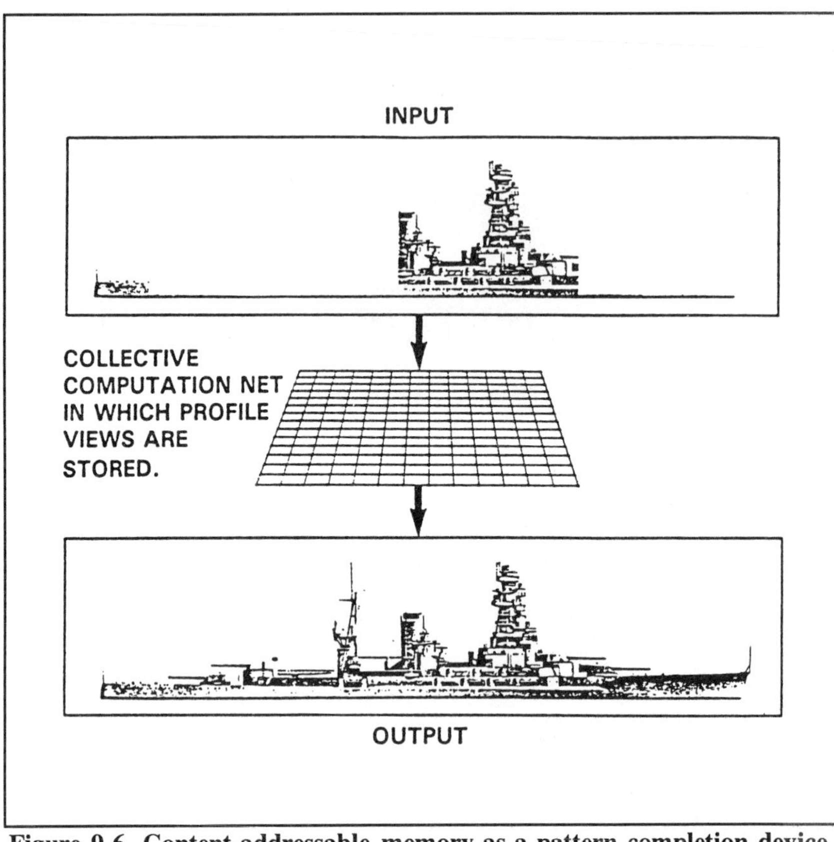

Figure 9.6 Content-addressable memory as a pattern completion device.

counterpart in the other. The number of these feed-forward connections bridging the two sides of the "retinotopic" structure is thus equal to the number of units in each. The two subnets could readily be exercised by two computers on two desks linked through the most ordinary local network. Only N bits need to be transmitted each KN^2t seconds, where K is the number of sweeps between updates and 1/t is the number of interconnects per second on each desktop.

Associative Memory Networks With Less Than Full Connectivity

When the network is used as a CAM in which certain subnets are clamped to represent input data, and the others are allowed to seek an energy

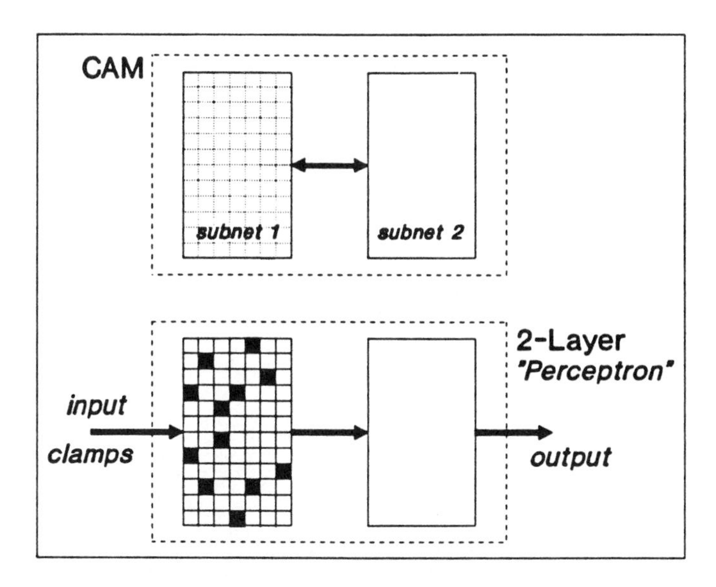

Figure 9.7 Clamping the units of one subnet gives rise to a "perceptron-like" network.

ninimum that completes the pattern, the full connectivity is effectively reduced. ᶠig. 9.7 (top) shows a CAM with two subnets of the same size that are fully reciprocally) interconnected, both internally and between each other. Treating he first subnet as input, clamping its N/2 units, the second subnet now receives ixed inputs through $(N/2)^2$ connections that feed forward. These can be ombined to form N/2 biases; and each sweep of the net can ignore the input inits. Were it not for the fact that the "output units" of subnet #2 are fully nterconnected, the network would reduce to the "two-layer perceptron" that has ome to be held in such universal contempt. Rosenblatt (1961,p. 477) saw the dvantage of back-coupling the output units as "the important tendency...to orrelate the output ... units so that they all apply to a single stimulus when a omposite stimulus occurs at the retina." In Hopfield's (1982) words, the strong ack-coupling will "produce categories, re-generate information and, with high robability, generate...the output O_1 from a confusing mixed stimulus $0.6O_1+0.4O_2$]."

In the next problem these back-coupling considerations do not apply, ince there is only one output unit.

Problem 9.1 (Prediction of Time Series): A process f(t) has the following history:

time (days ago), -t =	9	8	7	6	5	4	3	2	1	0
observation, f(t) =	1	0	1	1	0	1	0	1	1	0

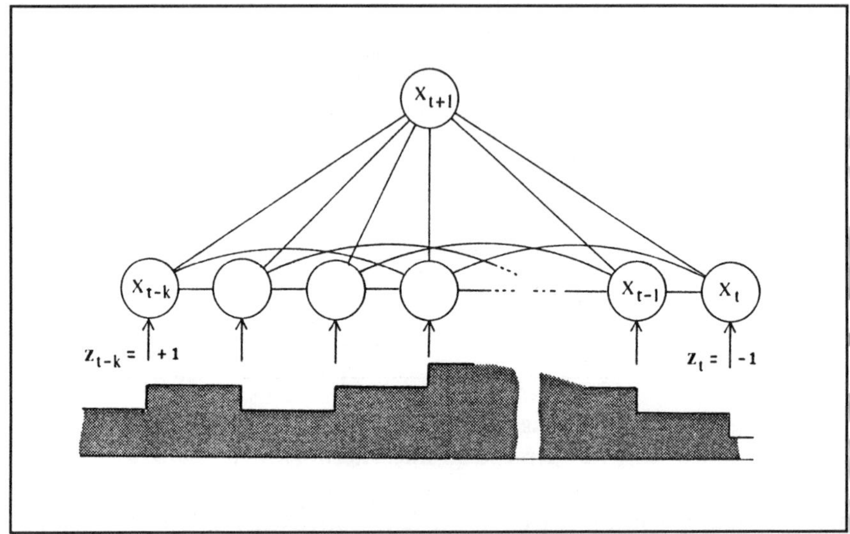

Figure 9.8 Prediction of time series.

Devise a collective computation network of five units to predict f(t+1) based on {f(t-k): k=0,1,2,3}.

Solution: Let the patterns be indexed by a=1,2,...,6. Each pattern has five (four inputs and one output) binary 0-1 variables: $x_i(a) = f(4 + a - i)$. Thus the pattern set is:

	x_1	x_2	x_3	x_4	x_5	
a = 1	1	0	1	1	0	
2	0	1	0	1	1	data
3	1	0	1	0	1	
4	1	1	0	1	0	...1011010110...
5	0	1	1	0	1	
6	1	0	1	1	0	t=-3

Note that the five-bit pattern $\mathbf{x}^{(4)}$ = (11010) represents the history of the process up to time t = -4 + 1 = -3. To store these patterns, first compute the pattern correlation matrix, the upper triangle of which is:

$$c = (1/6)\begin{bmatrix} 4 & 1 & 3 & 3 & 1 \\ & 3 & 1 & 2 & 2 \\ & & 4 & 2 & 2 \\ & & & 4 & 1 \\ & & & & 3 \end{bmatrix}.$$

(The remainder of the matrix can be completed from symmetry.) For example, c_{24} = (1/6)(2) = 1/3 because exactly two patterns (#2 and #4) have $x_2x_4 = 1$. Using the Hinton-Sejnowski formula:

$$W_{ij} = \ln\{[c_{ij}'(1+c_{ij}'-c_{ii}-c_{jj})]/[(c_{ii}-c_{ij}")(c_{jj}-c_{ij}")]\},$$

$c_{ij}'=c_{ij}+\epsilon$ and $c_{ij}"=c_{ij}-\epsilon$ for a suitably small ϵ, the weights are as shown in the network diagram below. If the time history now confronting the observer is:

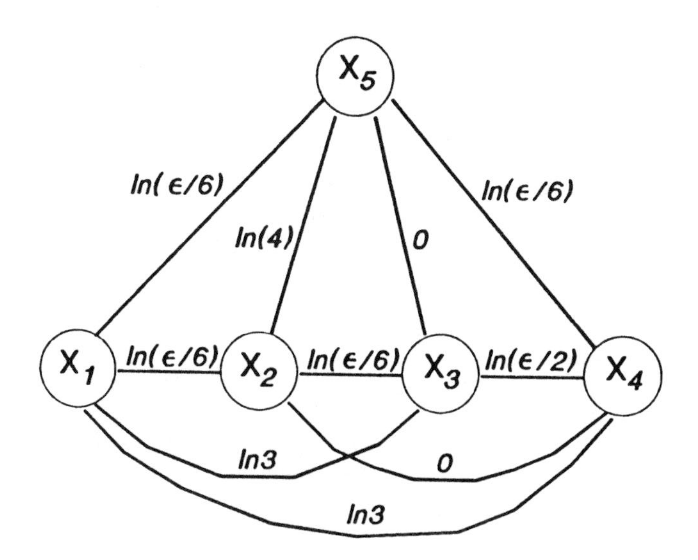

...1100, the input units #1 through #4 might be clamped so that $x_1=x_2=1$ and $x_3=x_4=0$. The effect of these clamps is to apply a net input of $\ln(4) + \ln(\epsilon/6) < 0$ to unit #5. So the prediction is $x_5 = 0$.

The clamping procedure amounts to creating a two-layer network with connections feeding forward from four input units to one output unit through weights W_{5j}, $j=1,\ldots,4$. The prediction is simply:

$$x_5 = \mathcal{H}[x_4 \ln(\epsilon/6) + x_3 0 + x_2 \ln 4 + x_1 \ln(\epsilon/6)]$$

or:

$$f(t+1) = \mathcal{H}\{f(t-2)\ln(4) + [f(t)+f(t-3)]\ln(\epsilon/6)\} \ .$$

Applying this predictor to the same data that generated the network, the prediction is correct five times out of six.

Problem 9.2: Sketch a collective computation network for "completing the pattern" represented by the daily history of the Dow Jones Industrial Average for the past twelve months.

Solution: Represent the quantized change in the price index from one close to the next by an integer $r \in \{-R,\ldots,-1,0,+1,\ldots, +R\}$. Let the patterns be the twelve-month histories with respect to 1,2,3,... days ago. The data representation features 2R+1 McCulloch-Pitts neurons for each of about 255 trading days of the year. With R=1, the network is a 3x255 strip of 765 units.

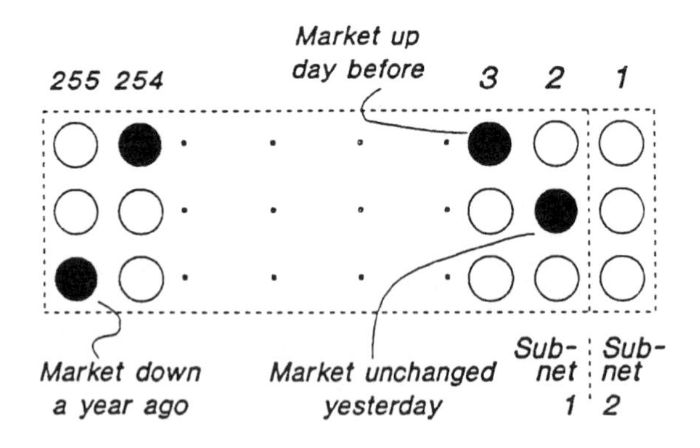

The pattern set could be derived from the past two years' daily closes, sufficient to produce 255 patterns of 765 bits each. A consequence of using the Hinton-Sejnowski formula to store the patterns is strong inhibition between units representing the change in the index on a particular day: If units i and j are in he same column (day), then $W_{ij} \propto \ln(\epsilon) << 0$. To predict today's close using he one year daily history, one could clamp the units in subnet #1 above and let he three units of subnet #2 reach a stable state in which, for ϵ sufficiently small, here will surely be an unequivocal (up, down, or no change) indication.

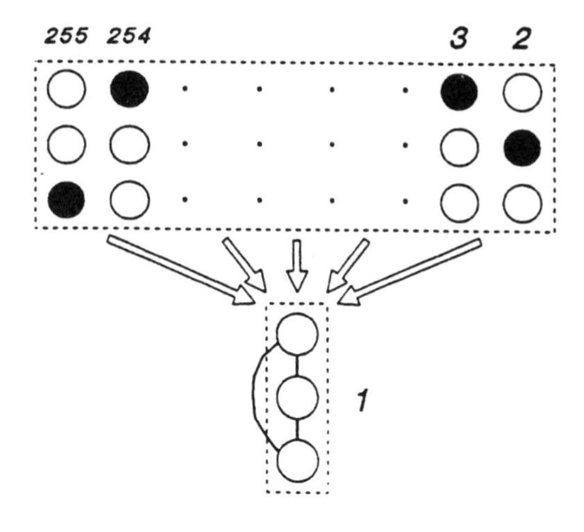

Then the network has effectively two layers with forward connections rom the clamped units to the three output units that compete in a winner-take-all ubnet.

Ludwig Boltzmann
(Photo courtesy of
The Ox Bow Press Woodbridge CT)

John Hopfield
(Photo courtesy of Dr.Hopfield)

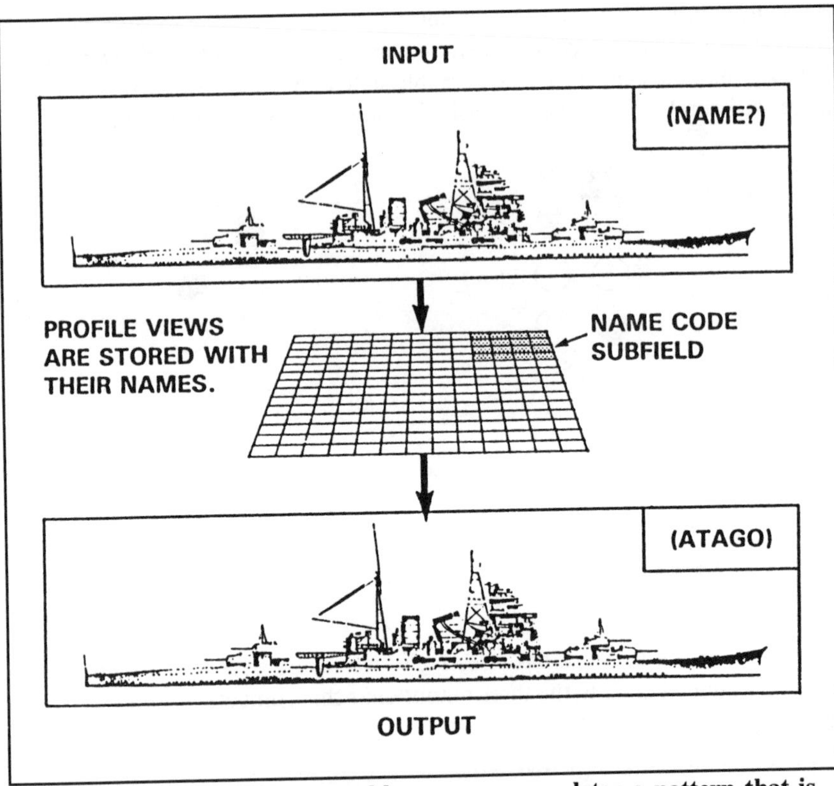

Figure 9.9 Content-addressable memory completes a pattern that is the union of two sub-patterns.

In fig. 9.8 there are again two subnets, one with 10^6+ units to represent the image of a warship and a second smaller subnet to display the name of the combatant. This imaginary network is created by scanning into the computer the pages of the U.S. Navy's recognition manual that show (and name) the ships of the Japanese Imperial Navy circa 1941. (If an optical disk can hold this much raw data, perhaps several of them can contain the weights computed with an appropriate storage prescription.) The CAM is tested by inputting the ship image. Pattern completion means forming the name in the subnet dedicated to this purpose. Since the ships are so few in relation to the pixels, it would not be unrealistic to expect that the energy landscape will be dominated by the fundamental memories. An input consisting of a partially blanked profile view would be unlikely to lead to the incorrect result indicated on the right side of fig. 9.8 unless the clamps are replaced by very weak biases.

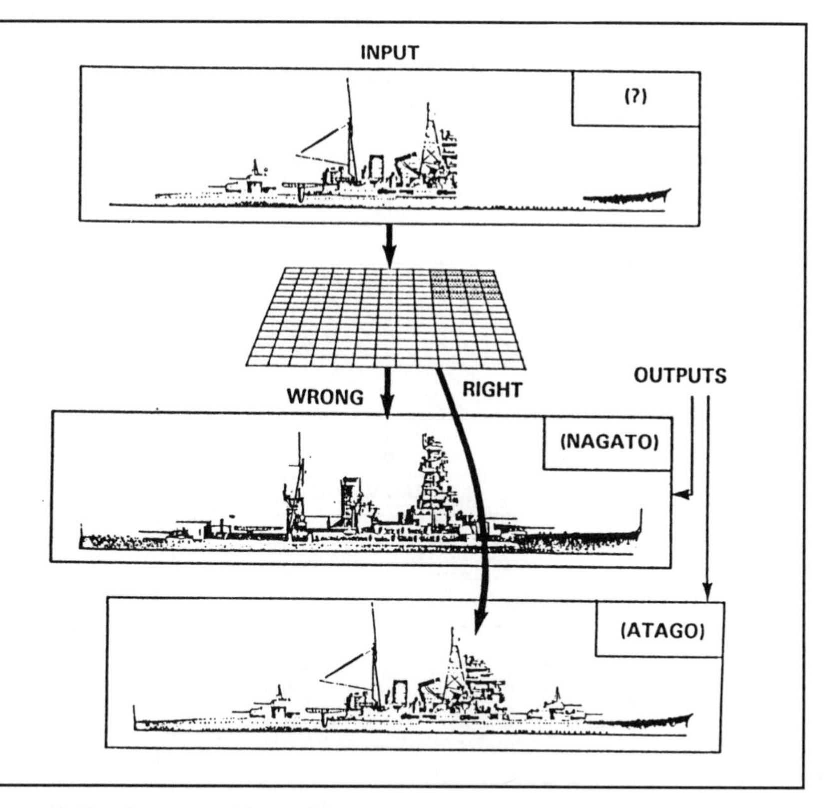

igure 9.10 Content-addressable memory completes a pattern that is the union of two sub-patterns.

In fig. 9.9 the profile views are stored with the aerial views; and a oudy view from the air leads to clear images from both perspectives. There not much realism in the way this CAM is exercised. As a result, clamped put units feed excitation and inhibition forward to output units the terconnections of which help to constrain the stable states to make them telligible.

Weakening the Clamps for Data Fusion

A more interesting mode of operation might consist in simultaneous puts from the air and from the surface, both too indistinct to yield an nmediate classification. The neural network associative memory would turally combine the evidence. If the two subnets are clamped according to the

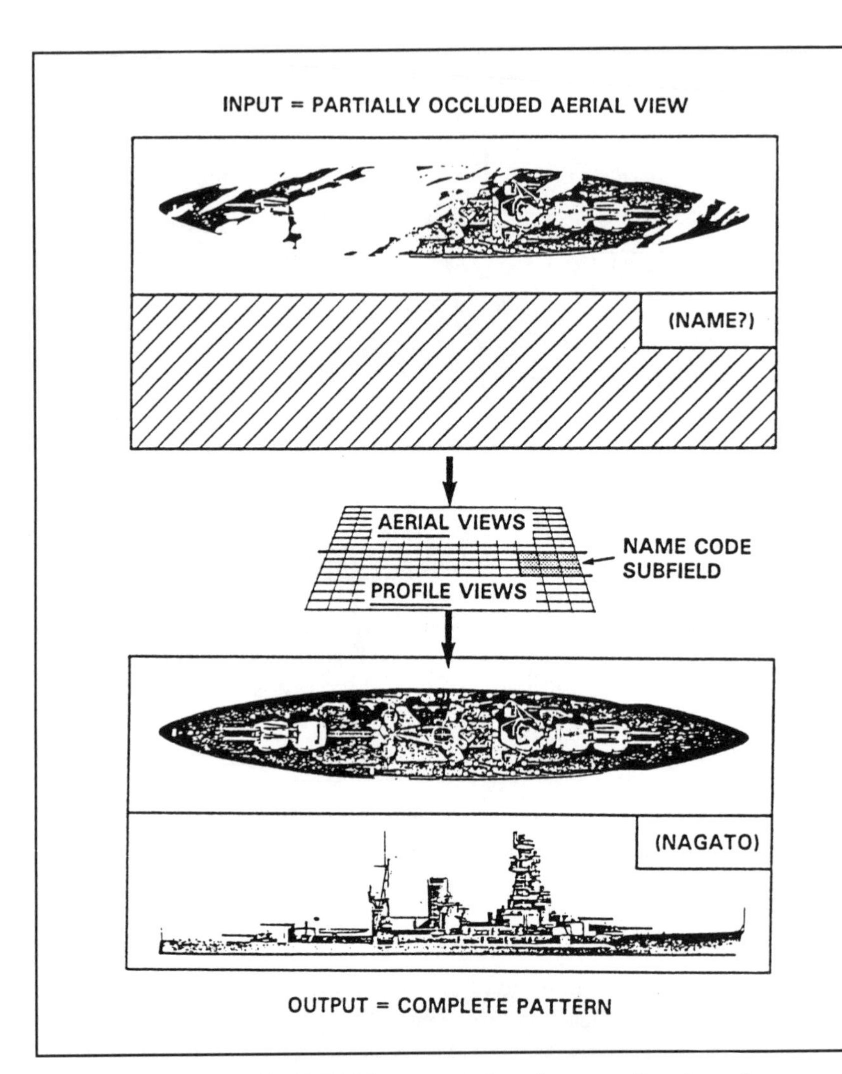

Figure 9.11 NNAM guesses view from surface based on imperfect aerial reconnaissance.

fuzzy data (the term fuzzy being understood in the colloquial sense), th resulting stable state of the name code subnet might be correct — to the exte that a two-layer "perceptron" with some back-coupling can do the job. Yet th uncertainty in the raw data seems to cry out for a gentler treatment. Instead c clamping the units in the visual subfields, let them receive biases that are stron enough to assert the sensory evidence but weak enough to permit the units t assume states which contradict their biases. In this way, a very crisp piece c one image can strongly influence the units in the fuzzy areas — and overrule the sensory data in these areas. Biases that are too strong in relation to the elements of the weight matrix could give rise to chimeras.

Little is known for sure regarding the correct way to bias the units. Referring to the second part of the Hinton-Sejnowski formula:

$$bias_i = z_i = -\ln(1/c_{i\,i} - 1)$$

in terms of the fraction $c_{i\,i}$ of the training patterns for which the ith unit is ON, Baran (1989) substituted for $c_{i\,i}$ the probability that the ith unit is ON in an ensemble of patterns obtained by adding noise to a signal (which may be one of the elements of the pattern set proper). If i is an input unit, let p_i be the likelihood that it would turn ON to represent a noiseless input of which the present input is a noisy realization. Then:

$$bias_i = -\sqrt{N}\,\ln(1/p_i - 1),$$

with the factor of $N^{1/2}$ to place the bias on the same order as the net input to unit i from the N-1 other units.

<u>Illustration</u>: In the stock market prediction problem (9.2), let the output (day 1) be tomorrow's close so that the right-most input represents the change on the present day (2). If it is twelve o'clock and the market is down, we might attribute probabilities of 0.2, 0.3, and 0.5 to the three possibilities (up, unchanged, down) of today's close (around 4 P.M.). Then the bias of the input unit representing a close **up** today is about:

$$-\sqrt{765}\;log(1/0.2 - 1) = -38$$

with reference to the last equation; and the bias of the unit representing a close **down** is <u>zero</u>.

The second author once stood before a conference on geophysics and remote sensing to describe a Markovian model for predicting the fluctuations in sea-surface roughness at a single location at six-hour intervals. The prediction was based solely on the past history of the process at the site in question. A critic in the audience pointed out that ocean waves are driven by the winds and that knowledge of the latter permits comparatively exact estimation of the former. Thus it seemed preposterous to look for information about the future in the past history of a process that is controlled by another measurable phenomenon. The fact that the Markovian model performed hardly better than the simple assumption that the sea state would stay the same from one observation to the next only highlighted the practical validity of the point made by the critic (who, it turned out, was one of the world's preeminent authorities on ocean waves).

Similarly, the economist will observe that the random process of stock-price fluctuations is influenced, if not controlled, by other variables (such as interest rates, exchange rates, tax laws, etc.). The predictive value of the information contained in the past history of the process could be practically nil. It would not be surprising to find that, if the network sketched above for stock-market prediction were actually constructed, the weights between units separated by more than a few days could approach zero [as $c_{ij} \approx c_{ii}c_{jj} \approx (1/2)^2 = 1/4$]. At the risk of belaboring the point made in the last chapter, storage prescriptions for the neural network associative memory are merely ways of capturing statistical correlations. If the "events" or "hypotheses" represented by the neurons are uncorrelated, the whole network resolves itself as a soup of unconnected subnets.

It may be (and probably is) impossible to predict the waves, the weather, or the Dow Jones Averages using only the past histories of such processes. Experts in these fields can develop mathematical models in which the sample path of the process to be predicted is influenced, either deterministically or stochastically, by other forms of "reconnaissance" or "side information". Or they may choose to give suitable, distributed representations to the process *and* the side information, so that these are represented in the subnets of a neural network associative memory. Such associative memories can be used to store the simultaneous histories of all the variables that influence the process and to discover statistical correlations among these variables (to the extent that they **are** correlated). When a stock-market prognosticator vouchsafes to us a (synthetic a priori) truth such as:

IF the 6-month trend is up
and IF interest rates are rising
and IF it is not an election year
and IF the moon is less than half full
THEN **sell everything**,

we are probably witnessing an *emergent property* of associative memory, namely the ability to combine diverse kinds of information with reference to an extremely large (but sometimes fuzzy) database called experience. When such rules fail to yield desired results, they tend to polarize the critics, some ontending that the wrong data representation was used in designing the ssociative memory, and others insisting that mistakes were made in the otoriously difficult process of "knowledge engineering".

10. Is There a Future in Collective Computation?

Seven years after the publication of "Neural Networks and Physic Systems with Emergent Collective Computational Abilities" (Hopfield, 1982 it is not uncommon to hear expert pronouncements to the effect that "th Hopfield model is dead". In October, 1989, the chief scientist of one of th three largest neurocomputer vendors told at least one audience that he did n know of *anyone* who was using the Hopfield network. Obviously the prese authors believe that these widespread obituary notices are somewhat prematu and exaggerated. (Mark Twain has asked that he not be quoted in this context At this point we hope the reader will agree, that it would be sad indeed for suc a fascinating mathematical structure, with such deep roots in the history physics and other fields, to be utterly without a worthwhile purpose in da processing of any kind.

We have mentioned the pioneering work of Rosenblatt several time It may be instructive to remember that, seven years after the appearance of (th Spartan Books edition of) *Principles of Neurodynamics: Perceptrons and th Theory of Brain Mechanisms* (Rosenblatt, 1962), the three-layer feed-forwa network of binary threshold units was deemed a dead end despite Rosenblatt description of a backpropagation technique for training the weights connections to the hidden layer. This assessment by influential experts with th most reputable institutions was based on grossly oversimplified models of th perceptron and their inherent limitations.

The fact that collective computation is today overshadowed (even to th point of being eclipsed) by other models may be irrelevant to its future exce in so far as these more popular approaches have all advanced on the coattails backpropagation, which took twenty years to rediscover. It may turn out th collective computation, like Rosenblatt's backpropagation, needs a very clev modification (and some time to sink in) before its practical implications can b fully appreciated by the industry.

The next paragraphs recall some of the critical points commonly made in regard to our subject; and we offer rebuttals to the extent that our experience in the preceding chapters permits.

1. "**The Hopfield net does not always converge to stable states**." This is certainly true when the synchronous dynamics is applied to symmetric networks of binary units. Synchronous updating often leads to persistent oscillations. As a biophysicist, Hopfield (1982) wrote that collective computation occurred *in spite of* asynchrony. It might have been more to the point, computationally, to say that energy minimization occurs *because* of asynchrony.

2. "**The Hopfield network converges to meaningless states** which correspond neither to (a) stored patterns (in the case of the CAM) nor to (b) legitimate solutions (in neural optimization circuits)." Point (a), which did not deter Hopfield, provides the inspiration for much ongoing research. Point (b) is symptomatic of the weakening of the "strong lateral inhibition" that enforces constraint satisfaction. This weakening, in turn, is probably motivated by the perhaps unrealistic expectation that "good, suboptimal solutions" must be found without the additive noise that permits the network state to escape from local energy minima.

3. "**The storage capacity of Hopfield's CAM is too low**; and, even if the patterns are orthogonalized before storage, equivalent performance can be achieved with bidirectional associative memories (BAMs) or two-layer nets that feed forward only." We expressed the view in chapter 8 that very little is really known about the capacity of the collective computation net as an associative memory. That the network can be trapped in (spurious) shallow minima at zero temperature is clear enough; and this motivated the development of the Boltzmann machine and the search for expedient procedures for temperature control. The capacity of the Boltzmann machine to find the desired patterns at deeper energy minima is essentially unexplored.

4. "**Convergence to optimal solutions requires simulated annealing procedures which are notoriously slow** and unsuited to most practical applications." The fact that the

logarithmic cooling schedule proposed by the Geman brothers is indeed too slow does not mean the problem has no expedient solution. While we remain skeptical concerning the fast simulated annealing procedures being offered at the present time, no proof of impossibility has been hinted at. The "computational thermodynamics" of neural networks seems capable of engaging the attention of many talented investigators for years to come.

5. "**The ability to recall stored patterns is insufficiently interesting** to warrant the assertion that collective computation networks will play a significant role in the development of artificial intelligence." Only time will tell if this is a valid objection. Ironically, it is sometimes voiced by those who, being enamored with the pattern-classification abilities of feed-forward networks, have never stopped to consider the functional equivalence of all neural network associative memories. The ability to "complete the pattern" makes the symmetrically interconnected CAM (Content Addressable Memory) a highly flexible component in many kinds of associative memory networks.

6. "**The Hopfield network has insufficient ability to generalize** beyond the horizons of the data and assumptions used in calculating the symmetric weight matrix. **It lacks the ability to learn and to self-organize** for the solution of practical problems in which little is known a priori about the environment (or stimulus world) of the network." Farley (1960) observed that one of the main problems in constructing connectionist models of learned perception is to provide a suitable means of "generalization" from particular stimuli to new stimuli never before encountered. In his view, the physical realization of a solution to the problem of generalization should "possess very nonlinear static and dynamic characteristics", having "stable points" in a space of "property classes" to which nearby trajectories are attracted. "The nonlinear action may then result in a locking-in of input to class. A one-dimensional analogy in engineering might be the lock-in action of the AFC in a radio receiver." Twenty years later, collective computation provided a rather compelling N-dimensional analogy to the mechanism that Farley was looking for. In another ten years it may be

possible to design neural network associative memories in which Farley's property classes can be realized in the state space of the network and perhaps identified with the substates of particular subnets for the detection and characterization of "key features".

7. "**Full connectivity is impractical** as the storage of N^2 weights for N units uses too much RAM in a digital simulation model and too much circuitry in analog VLSI." This is probably true for N large enough to represent most practical problems. We can only hope that associative memory networks can be designed with a little (human) ingenuity so that the data representation supports a natural partition of the whole into subnets that are fully interconnected within themselves but that communicate by less than full connectivity with the units of the other subnets. This ingenuity in data representation together with (next-generation?) hardware for concurrently asynchronous processing — either digitally with local memory multiprocessors or in multiplexed analog circuits — could make its first appearance in connectionist models in which symmetrically interconnected subnets form layers between which connections feed in primarily one direction. Generalizations of backpropagation to recurrent networks might make them so popular as to abolish the schism between structural types that so dominates the scene today. An innovative researcher might even (today) train a backpropagation network to classify all the local minima of a Hopfield net as belonging to one of the fundamental memories. The hard part would come in the analysis of the successful experiment: Did the "generalization" occur in the training of the forward connections or in the energy landscape of the symmetric subnet?

If collective computation survives the pessimism of the moment we may look forward to innovative new network architectures which (like the cerebral cortex) combine subassemblies in which connectivity takes radical and unique forms. Moreover, its main results are really all one has to go by in seeking the principles that would make more general kinds of recurrent (back-coupled) networks understandable. In this spirit we are confident that the most important and interesting problems in collective computation have yet to be formulated.

Appendix A: Convergence in Time in Asymmetric Neural Networks

Introduction

It can be argued that real neurons (unlike Ising spins) are not bistatic and must be represented by continuous systems of coupled, nonlinear differential equations; and that their interactions involve temporal relationships, like propagation delays and refractory behavior, that have yet to be captured by any system of equations. Thus Pineda (1987) argues that the precise definition of neurodynamics remains elusive. Yet even the simplistic models can exhibit interesting behavior, particularly their convergence to stable states corresponding to memories which are addressable by content instead of location. Networks having asymmetric connectivity have received recent attention, since symmetry has been so widely criticized as an artifice in the best known models, like those of Cohen and Grossberg (1983) and Hopfield (1984). Do random patterns of excitatory (positive) and inhibitory (negative) weights support global pattern formation? Pineda says they do while Lapedes and Farber (1986) assert the opposite, at least in the case of continuous systems.

Glauber Dynamics

Hopfield (1982) employed a stochastic dynamics in which each neuronlike unit is interrogated at the same mean rate and updated according to the decision rule of the (0-1 binary) McCulloch-Pitts neuron. The states of this system are the vertices of a unit hypercube and the motion is in jumps between adjacent vertices, since the units are interrogated in a random serial order. The stochastic trajectory followed by the network state vector is a continuous time Markov chain as posited by Glauber (1963), who introduced essentially the same dynamics (at finite reciprocal temperature) to study the Ising model (cf. Cipra 1987). A discrete time Markov chain is imbedded in the process by ignoring the durations of the intervals between interrogation steps. The dynamical equation is:

$$x_i(t+1) = \begin{cases} \mathcal{H}\left[\sum_{j=1}^{N} W_{ij} x_i(t)\right] \ \textit{if} \ \ i = K(t) \\ x_i(t) \qquad\qquad \textit{otherwise} \end{cases} \tag{1}$$

where $\{K(t),\ t = 0,1,2,\ldots\}$ is a sequence of independent random numbers uniformly distributed among the first N integers that index the McCulloch-Pitts neurons). Assume $\mathbf{W}_{ii} = 0$. Spin variables $2x_i - 1$ can be substituted for the indicatory variables, if the sign function replaces the unit step $\mathcal{H}(\cdot)$; and the weights $\mathbf{W}_{ij}$ will then, by convention, be denoted $\mathbf{J}_{ij}$ as in most discussions of spin systems. Crisanti and Sompolinsky (1988), after writing the dynamical equation in this latter way, conducted an extensive Monte Carlo study of convergence time in networks with varying degrees of asymmetry. The following points will clarify their findings.

Transition Matrix

The $2^N \times 2^N$ stochastic (one-step transition) matrix **P** of the Markov chain is:

$$P_{mn} = Pr\ [x(t+1) = x_n | x(t) = x_m]$$

where subscripts m and n index the states. (The canonical indexing scheme would be $m(\mathbf{x}) = \Sigma_i\, x_i\, 2^i$.) Let d_{mn} be a vector of N binary digits the ith of which, d^i_{mn}, is equal to $|x_{ni} - x_{mi}|$. The Hamming[1] distance D(m,n) is the number of nonzero components of this vector. Equation (1) allows only transitions from a vector to a neighboring vector — that is, one whose Hamming distance, D(m,n), from the vector is unity.

$$P_{mn} = 0 \quad \textit{if} \quad D(m,n) > 1 \ . \tag{2a}$$

The argument of the step function in (1) is the ith element of the (column) vector $\mathbf{Wx_m}^\dagger$, where the dagger denotes transposition (and **x** is defined as a row

[1] See page 46.

vector). If a transition of distance one occurs with $\mathbf{x}_n$ as the final state of the interrogated unit, then[2] $\mathcal{H}(\mathbf{d}_{mn}\mathbf{W}\mathbf{x}_m{}^{\dagger}) = \mathbf{x}_n$. Given an $x_{\mathbf{n}}$ for which D(m,n) = 1, the quantity:

$$\Delta_{mn} = \begin{cases} \mathcal{H}(d_{mn}Wx_m^{\dagger}) & \text{if } \; x_m d_{mn}^{\dagger} = 0 \\ 1 - \mathcal{H}(d_{mn}Wx_m^{\dagger}) & \text{if } \; x_m d_{mn}^{\dagger} = 1 \end{cases}$$

is unity whenever the one step transition from the mth state to the nth state can occur. Thus:

$$P_{mn} = (1/N)\Delta_{mn} \quad if \quad D(m,n) = 1 \quad . \tag{2b}$$

with the factor of 1/N being the probability that the unit in question is interrogated at the tth step. Finally:

$$P_{mm} = 1 - \sum_{n \neq m} P_{mn} \tag{2c}$$

along the diagonal. Equations (2) fully specify the transition matrix.

The fixed points $\{\mathbf{x}^0\}$, which satisfy:

$$x_i^0 = \mathcal{H}(\sum_{j=1}^{N} W_{ij}\, x_j^0) \quad ,$$

i, j = 1, . . ., N, are now "absorbing states"[3] $\{\mathbf{x}^0\} = \{\mathbf{x}_m | P_{mn} = 1\}$ and following Feller (1957), each state is classified as transient or persistent. The closure C of a persistent state is an irreducible set and its submatrix defines a Markov chain on it which can be treated independently of the rest. Consider for example a network of four neurons, depicted in Figure A.1, with:

---

[2] Note that $\mathbf{d}_{mn}$ has only one nonzero component $\mathbf{d}_{mn}\mathbf{W}\mathbf{x}_m{}^{\dagger}$ is equal to the corresponding component of $\mathbf{W}\mathbf{x}_m$.

[3] See page 91.

$$W = \begin{bmatrix} 0 & 1 & 0 & 0 \\ +1 & 0 & 0 & 0 \\ 0 & -1 & 0 & +2 \\ 0 & +1 & -2 & 0 \end{bmatrix}$$

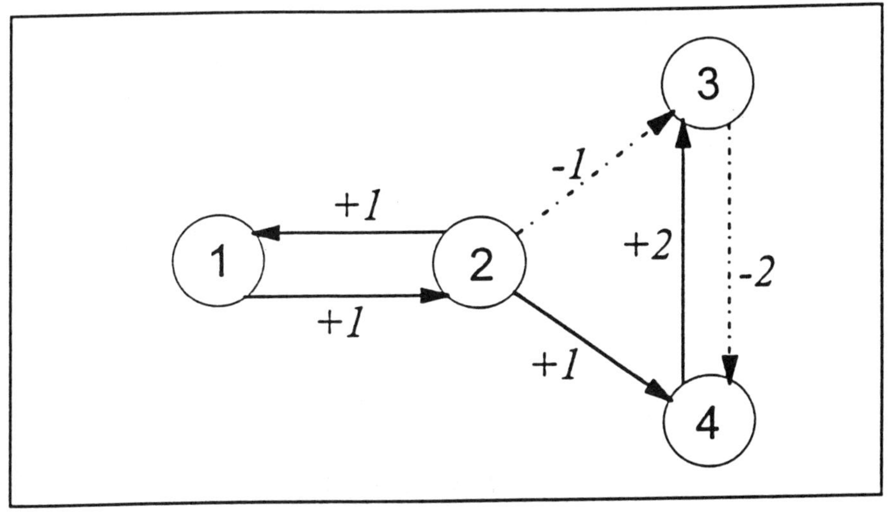

Figure A.1 A network of four McCulloch-Pitts neurons which exhibits astable limiting behavior.

Although **P** is 16 x 16, a moment's reflection shows at least one absorbing state $\mathbf{x}^0 = (0000)$ and at least one irreducible set C_i for which the lower-right 4×4 submatrix of the incomplete array is shown below. The columns, indexed by the 4 vector $\mathbf{x}_n$ using $n = \sum_{i=1}^{4} x_i 2^i$, give the transition probabilities. Starting from $\mathbf{x}(0) = (1111)$, the limit cycle for C_1 is $1110 \rightarrow 1100 \rightarrow 1101 \rightarrow 1111$. This is not a periodic process, since the Markov chain is characterized by geometric holding times. In fact, the zero state and C_1 absorb $\mathbf{x}(t)$ for every $\mathbf{x}(0)$. One thousand trials led from random $\mathbf{x}(0)$ to the origin 560 times and to C_1 in all the rest. Small networks in which nonzero stable states coexist with irreducible sets of greater cardinality can also be devised.

x_m \ x_n	0000	. . .	1100	1101	1110	1111
0000	1	. . .				
.	.	.				
.	.	.				
.	.	.				
1100			3/4	1/4	0	0
1101			0	3/4	0	1/4
1110			1/4	0	3/4	0
1111			0	0	1/4	3/4

Monte Carlo Calculations

To estimate the likelihood of reaching a stable state in a network of given type, the obvious recourse is to Monte Carlo calculations with the following steps:

(a) Create a **W** (N × N) with some stochastic rule.
(b) Create an **x**(0) of N independent binary components that are 0 or 1 equiprobably.
(c) Iterate equation (1) until an $\mathbf{x}^0$ is reached, or until t_{max} iterations have failed to find one.
(d) Repeat these steps a large number of times for each N.

The time limit t_{max} is conservatively estimated as $N2^N$, since the state space could be searched exhaustively in about this time. Hopfield observed convergence in most symmetric nets in about 4N iterations, corresponding to four time constants of the analog circuit model. Thus astable behavior after $N2^N$ iterations, or 2^N random sweeps of the net, would signify analog circuit instability outlasting 2^N time constants. There is another reason to consider this selection. Imagine a transition matrix, **P**, in which Q states would form an irreducible persistent set C in the absence of $\mathbf{x}^0$. But $\mathbf{x}^0$ is accessible from a single $\mathbf{x}^*$ in C. This $\mathbf{x}^*$ would have a mean recurrence time, t^*, in the absence of $\mathbf{x}^0$, but the mean time to absorption into $\mathbf{x}^0$, after C is entered, will be approximately t^*m, where m is the mean number of times $\mathbf{x}^*$ is visited by $\mathbf{x}(t)$ before absorption. The mean number of visits to $\mathbf{x}^*$ is one more than the mean number of escapes, which equals the number of states (in C) accessible from $\mathbf{x}^*$. If each state in C can lead to only one other (as in the simple example above), then $m = 2$ and $t^* = NQ$. Then the mean time to absorption is upper bounded by $2N(2^N - 1)$, since, in the worst case, the entire state space (except $\mathbf{x}^0$) belongs to C.

Figure A.2, a normalized, accumulated histogram of forty convergence times with N = 9, is typical of the results obtained using the above procedure, taking uncorrelated, zero-mean, logistic random variables for the off-diagonal elements of $\mathbf{w}$. Though the abscissa is logarithmic, the upper tail of distribution is extremely heavy. With 100 trials for each N, use of $t_{max} = N2^N$ gave 100-B(N) values of the convergence time, averaged under T in table A.1. There is no obvious trend in the numbers B(N) of nonconvergent trajectories. These crude results differ only superficially from the findings of Crisanti and Sompolinsky (1988), whose Cray X-MP simulations showed almost no cases of convergence in thousands of randomly generated asymmetric networks with N > 100. They imposed a time limit of $t_{max} \approx$ 15,000 Monte Carlo steps per neuron (spin), slightly (!) fewer than the 2^{14} that would be required to extend table A.1 to N = 14.

Table A.1 Mean convergence time (T) and number of nonconvergent trajectories (B) for indicated numbers of neurons (N), based on 100 independent trials at each N.

N	T/N	BNT/NB
4	1.8	3912.29
5	2.9	11050.810
6	4.1	51141.28
7	5.3	712100.55
8	18.9	1313179.06

Figure A.3 highlights the extent to which the symmetry of the connectivity matrix expedites convergence. The curve to the right was obtained by plotting and connecting 5k-percentile points (k = 0,1,2,...) of the

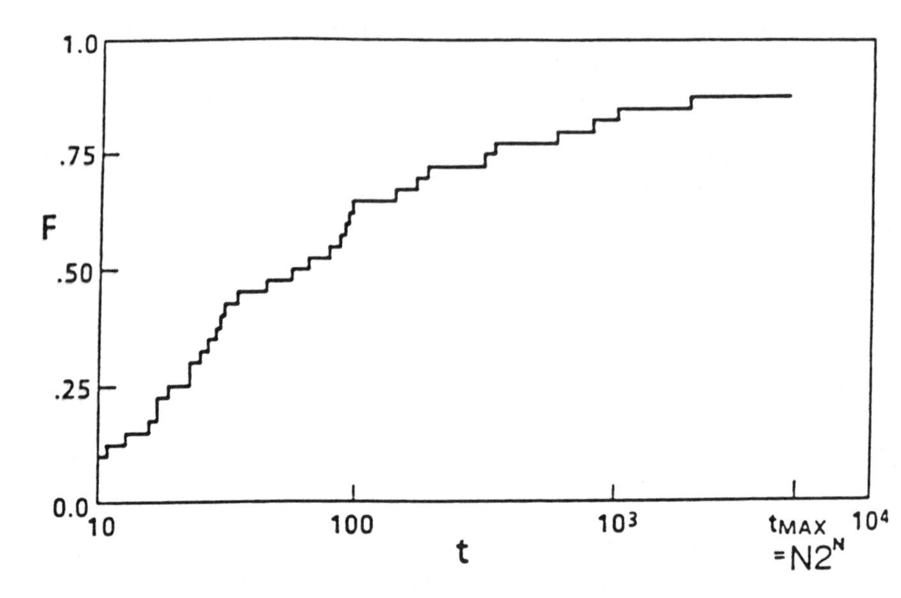

Figure A.2 Sample distribution function (F) of the convergence time (*t*) in 40 random networks of size *N* = 9.

Here F(t) is the fraction of specimen nets in which the state converges to a fixed point before t + 1 iterations.

the same procedure described above. The curve to the left shows the distribution of the convergence time in 100 runs with symmetric networks of the same size. In this case the abscissa can be called the relaxation time (RT), since it corresponds to the stochastic relaxation of a global energy function. Both sets of trials used logistic random variables for the weights, with $W_{ij} = W_{ji}$ in the case of RT, but the two are statistically independent in case CT. Crisanti and Sompolinsky (1988) have attributed a superficially log-normal distribution to the convergence time in partially asymmetric nets, asserting (equation 6.5) that the mean of ln(CT) can be written as $Na(\kappa)$ in terms of an index of asymmetry κ, which ranges from zero (in perfectly symmetric nets) to one (in fully asymmetric ones). Moreover, they state that $a(\kappa) \rightarrow 0$ as $\kappa \rightarrow 0$, and $a(\kappa) \rightarrow \infty$ as $\kappa \rightarrow 1$. These conclusions may be artifacts of the time limit (15,000 N). The personal work of one of the authors suggests that, in symmetric networks, ln(RT) is approximately logistic; and the median RT exceeds $(N/2)^{3/2}(\ln N)^{1/2}$ iterations.

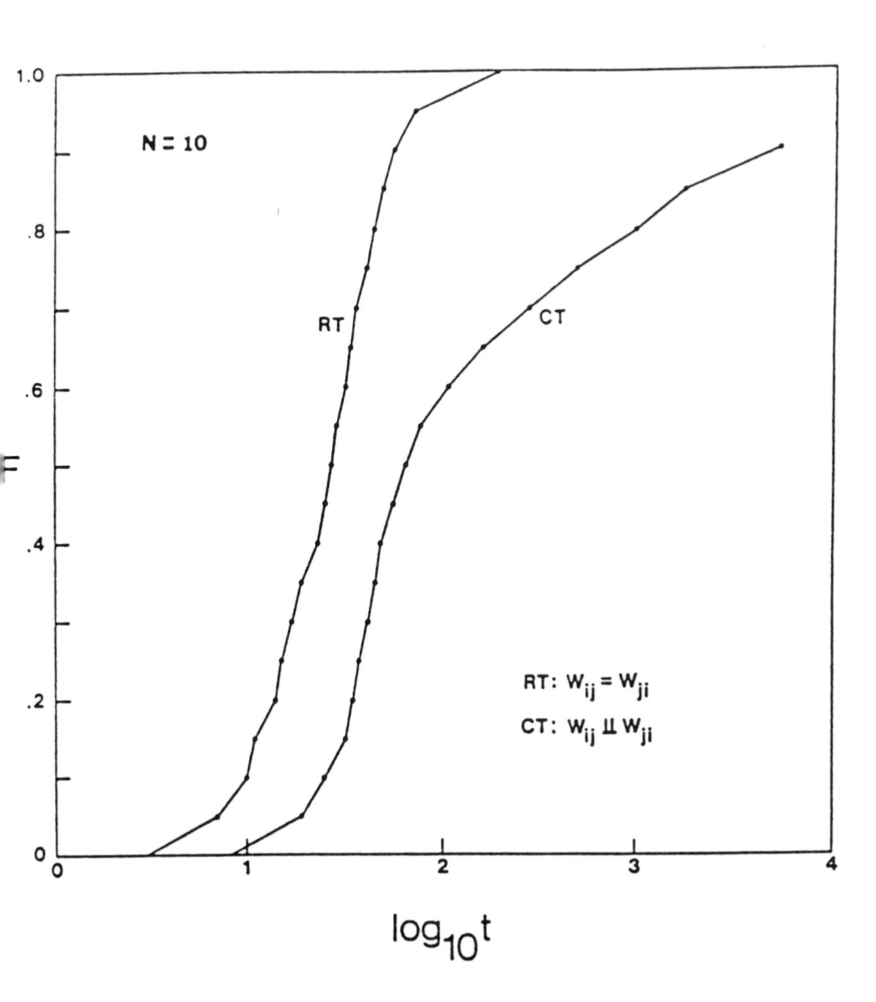

Figure A.3 Relaxation time (RT) in symmetric nets and convergence time (CT) in asymmetric nets: cumulative distribution function (F) of the time (t) in iterations.

Conclusions

It seems that Monte Carlo calculation of mean convergence time will be inherently difficult when fully asymmetric networks are considered. For N = 100, 10^{34} inner-product steps would be required to reach $t_{max} = N2^N$. While there is no guarantee that this time limit is sufficient for large N, we can be certain that practical work in neuromorphic cybernetics will be restricted to network architectures that for the most part converge more rapidly than the randomly configured.

Appendix B: Markov Chains Arising in Collective Computation Networks with Additive Noise

Introduction

One of the folk theorems of statistical neurodynamics holds that the globally asymptotically stable states of a neural network occur with relative frequencies given by the Boltzmann distribution when the network is subjected to isothermal agitation. Hopfield (1982) showed that the states had an associated computational energy function:

$$H(s) = -\sum_{j=i+1}^{N} \sum_{i=1}^{N-1} W_{ij} \, x_i x_j - \sum_{i=1}^{N} z_i \, x_i \tag{1}$$

which, as Lyapunov has proven, guarantees global asymptotic stability. The computational energy of the network is analogous to the Hamiltonian of a collection of strongly interacting magnetic dipoles (the Ising spins).

For binary McCulloch-Pitts neurons:

$$x_i = \begin{cases} 1 \text{ if } u_i > 0 \\ 0 \text{ if } u_i \leq 0 \end{cases} \tag{2}$$

when the W-matrix is real-valued and symmetric, with all zeros on the diagonal, the network evolves toward stable states that correspond to local minima of the computational energy. (See chapter 3.) The "energy landscape" can be configured so that these local minima correspond to solutions of constrained optimization and pattern recognition problems. (See chapter 5.)

"Boltzmann Machines"

Ackley, Hinton, and Sejnowski (1985) proposed simulated annealing to dislodge the Hopfield network from local minima and enable it to settle into states of still lower energy that would represent better (if still suboptimal) solutions. (See chapter 7.) The network is "heated" by the addition of noise to the input of each unit. When these noises are independent, identically distributed random variables, the state **x** takes a random walk on the 2^N vertices of a hypercube. The stationary distribution is:

$$Pr(\boldsymbol{X} = \boldsymbol{x}) = \frac{\exp[-\beta H(\boldsymbol{x})]}{\sum_{\boldsymbol{x}'} \exp[-\beta H(\boldsymbol{x}')]} \quad . \tag{3}$$

The assertion of Ackley, Hinton, and Sejnowski, that $1/\beta = T$ is proportional to the root mean intensity of noise described by a logistic distribution, was not powerfully motivated. Shaw and Roney (1979) had earlier arrived at an expression like (3) in which β is a "smearing factor" determined from details of a stochastic model of the chemical synapse.

The Algorithm

The computational technique of simulated annealing traces its roots to the Metropolis (1953) algorithm, which updates the state of an N-particle system according to a stochastic model in which the Boltzmann distribution is expressly assumed beforehand. Our research has led to an alternative derivation. The stochastic model leads to a simple algorithm as follows. Pick an integer i at random from the set {1,...,N}. Select a real random variable, call it Z_i, which is symmetrically distributed about a mean of zero. Compute:

$$x_i = \mathcal{H}\left(\sum_{j=1}^{N} W_{ij}x_j + Z_i\right).$$

These steps are iterated indefinitely with independent, identically distributed random numbers (Z_k, k = 1, 2, 3,...). It is not hard to see that in the case of the Glauber dynamics, this gives rise to a sequence (s_k, k = 1,2,...) of states that constitute a Markov chain. Nonzero probabilities can be attributed to transitions that involve at most one component of the state vector. With no external input ($\mathbf{z} = \mathbf{0}$), these probabilities depend on the W_{ij} and the distribution of the resulting $\mathbf{Y} = \mathbf{Wx}$. When Hopfield's conditions (theorem 3.1) are obeyed by the former, the stationary distribution can be derived analytically. This distribution is:

$$Pr(x=x) = z^{-1}\exp\left(\sum_{j=i+1}^{N}\sum_{i=1}^{N-1} x_i x_j \, ln \, \frac{F(W_{ij})}{1-F(W_{ij})}\right) \tag{4}$$

in which F is the (cumulative) distribution function of Y and the denominator Z is the sum over all states which normalizes the discrete probability density. The assumption of logistic noise, as:

$$F(y) = \frac{1}{1+e^{-\beta y}}, \quad -\infty < y < \infty \quad , \tag{5}$$

gives the last equation a particularly simple form (3).

Transition Matrix

To derive the transition matrix of the Markov chain $\{\mathbf{x}_k, k = 1,2,\ldots\}$, let $\mathbf{x}$ and $\mathbf{x}^\dagger$ be the network state as a column and row matrix, respectively. Let $\mathbf{d}_j$ denote a column vector that has N components the ith of which is δ_{ij} (the Kronecker delta). Consider just the case of no input ($\mathbf{z} = \mathbf{0}$). Then $u_j = \mathbf{x}^t\mathbf{W}\mathbf{d}_j$ where $\mathbf{W}$ is the weight matrix subject to Hopfield's restrictions (theorem 3.1). The Glauber algorithm selects a j at random and computes $x_j = \mathcal{H}[u_j + y]$, where $\mathcal{H}[.]$ is the unit step and y has distribution function, F(y), whose density function, f(y), is symmetric about $y = 0$.

We want the probability of a transition from state $\mathbf{x}$ to state $\mathbf{x} + \mathrm{d}\mathbf{x}$, where $\mathrm{d}\mathbf{x} = \mathbf{d}_j$ as defined above if $x_j = 0$ and $\mathrm{d}\mathbf{x} = -\mathbf{d}_j$ if $x_j = 1$. This probability, denoted $Q(\mathbf{x} + \mathrm{d}\mathbf{x}|\mathbf{x})$, is proportional to 1/N, the probability that j is selected, and is given by:

$$Q(x+dx|x) = \begin{cases} (1/N)\; Pr\{Y+U_j > 0\} & \text{if } dx = d_j \\ (1/N)\; Pr\{Y+U_j \le 0\} & \text{if } dx = -d_j \end{cases}$$

in which U_j is determined by $\mathbf{x}$ as noted. These statements are the same as:

$$Q(x+dx|x) = \begin{cases} (1/N)\ F(x^{\dagger}Wd_j) & if\ dx = d_j \\ (1/N)\ [1-F(x^{\dagger}Wd_j)] & if\ dx = -d_j \end{cases}$$

because of the symmetry of the distribution of Y. For transitions of zero Hamming distance we shall have:

$$Q(x|x) = 1-\sum_{dx} Q(x+dx|x) \quad .$$

For transitions of distance more than one, the probability is zero, since the algorithm specifies the interrogation of the units one at a time.

Asymptotic Temperature

With regard to equation (4), suppose that the root mean intensity of the noise is large compared to each $W_{ij} = t/\beta$. Then the first-order Taylor series expansion of the logarithm is:

$$\log\frac{F(t)}{1-F(t)} \approx 4tF'(0)$$

since $F(0) = 1/2$. Defining the asymptotic temperature T_∞ of the network in such a manner that $\beta = 1/T_\infty$ in (3), we shall have:

$$T_\infty = 1/[4f(0)] \qquad \textbf{(6)}$$

in terms of the probability density $f(y) = F'(y)$. When (5) is assumed, the last equation is indeed valid for $\beta = 1/T$. If the noise were normally distributed with standard deviation σ, then the asymptotic temperature would be $T_{NORMAL} = (\frac{1}{2}\pi\sigma)^{\frac{1}{2}}/2$. If the noise has a Cauchy density $f(y) = (c/\pi)/(c^2 + y^2)$, the temperature would be $T_{CAUCHY} = \pi c/4$. Clearly this asymptotic temperature is not a function of the mean noise intensity, since the variance of the Cauchy random variable is undefined.

Simulations

Figure B.1 represents a Hopfield net of four units in which the labeled segments give the dimensionless strengths of the symmetric interconnections. Let the inputs to units #1 and #2 be denoted A and B, respectively; and let

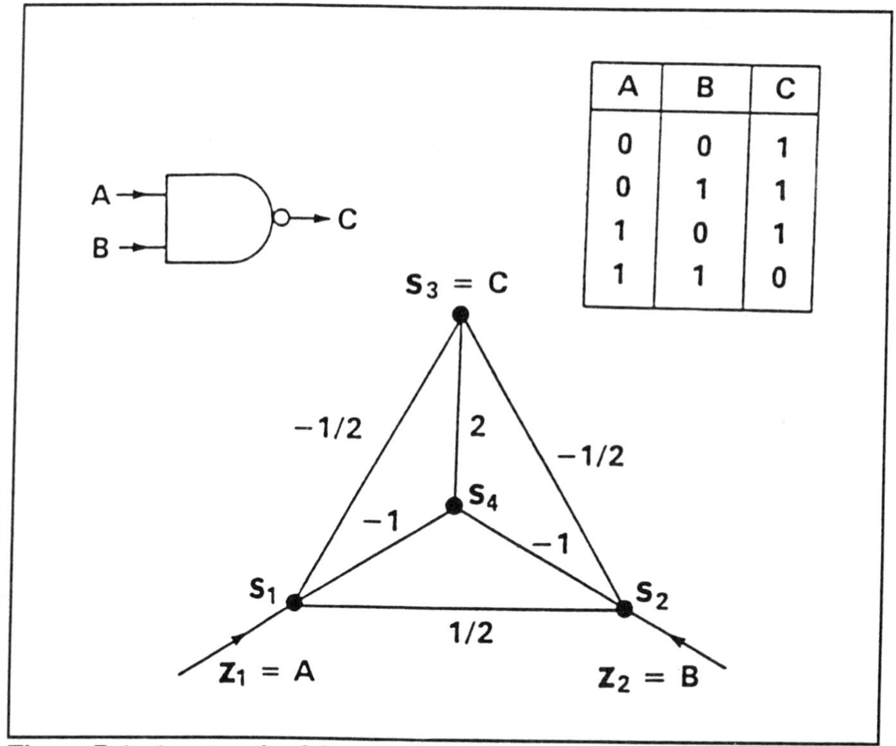

Figure B.1 A network of four units (nodes), two of which receive binary inputs (A & B) and one of which registers the output (C). The connectivity matrix is specified by the labels on the line segments linking the units.

s_3= C. We shall consider only binary {0,1} inputs. The weights are chosen so that this small network performs the NAND (Not-AND) logic function C(AB) which the truth table (in the upper right-hand corner of the figure) defines. This would indeed be the case if the network always settled into the state that gives the global or absolute minimum energy. Table B.1 uses the formula:

$$m(s) = \sum_{i=1}^{4} s_i 2^{i-1}$$

to assign a natural number m to each of the sixteen states of the network; and it lists the negatives of the energies of the states for each input condition AB $\in$ {00,01,10,11}. With input AB = 11, the minimum energy is -2.5 and it occurs in state m = 3, for which C is zero. With the other inputs, the minimum energy is -2.0 and occurs in state m = 12, for which C = 1. This motivates the truth table of Figure B.1.

Table B.1 Negative energy values of the sixteen states of the four-unit NAND gate.

	AB			
STATE:	00	10	01	11
0	0	0	0	0
1	0	1	0	1
2	0	0	1	1
3	-.5	1.5	1.5	2.5
4	0	0	0	0
5	-.5	.5	-.5	.5
6	-.5	-.5	.5	.5
7	-.5	.5	.5	1.5
8	0	0	0	0
9	-.5	0	-1	0
10	-.5	-1	0	0
11	-1.5	-.5	-.5	.5
12	2	2	2	2
13	.5	1.5	.5	1.5
14	.5	.5	1.5	1.5
15	-.5	.5	.5	1.5

Figure B.2 is a state transition map to show which transitions are allowed. Since units are interrogated in a random serial order, only one unit can toggle at a time. Thus the allowed transitions are of Hamming distance one. The sixteen states of the "NAND gate" correspond to squares in the 4×4 array of the map. The squares are labeled with the values $m(\mathbf{x})$. Motion is horizontal or vertical — never diagonal — between adjacent squares. The map wraps around horizontally and vertically as indicated by the connecting lines and arrows.

The negative energy map of Figure B.3 consists of four submaps each with the structure of the preceding figure. Here each square is labeled with -

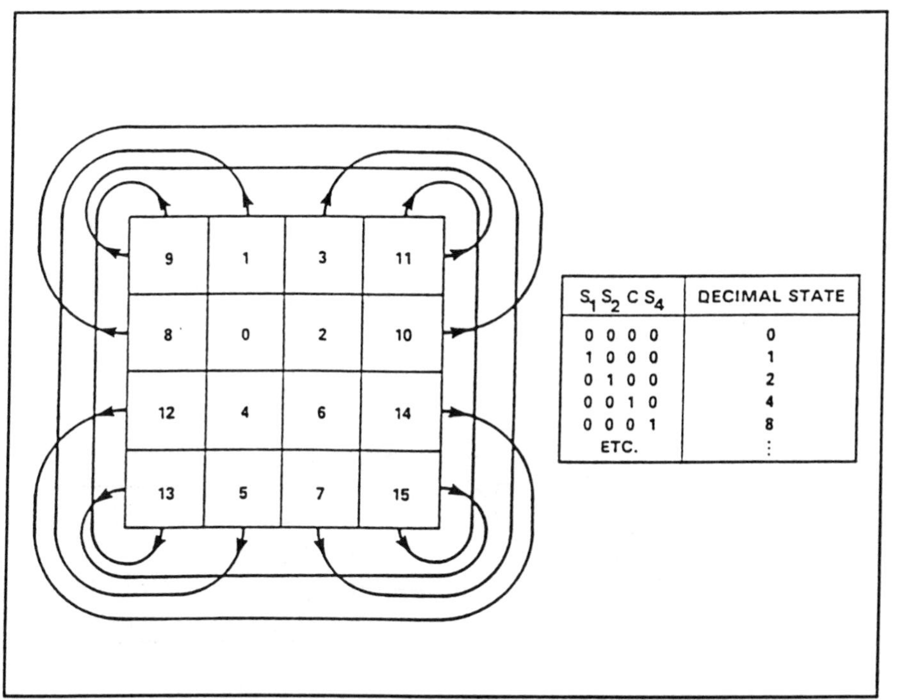

S_1 S_2 C S_4	DECIMAL STATE
0 0 0 0	0
1 0 0 0	1
0 1 0 0	2
0 0 1 0	4
0 0 0 1	8
ETC.	⋮

Figure B.2 The state transition map for the network of Figure 1 uses the indicated binary-to-decimal convention to assign an integer (0 through 15) to each state of the network.

$H_m(AB)$ where A and B are the strengths of the biases keeping z_1 and z_2 (respectively) on, and m is the index of the state vector, **x**: $m = \sum_i x_i 2^i$. The four submaps correspond to the four input conditions. If the network begins in state m = 3 with AB = 11, the energy is minimized and the state is stable. Now if the input changes, the network is unable to leave the initial state, because any allowed transition will increase the energy. Similarly, if the initial state is m = 12, and the input is subsequently set to AB = 11, the state cannot assume the desired value (m = 3) except by way of intermediate states of higher energy.

When noise is injected into the units of the network, the state can be dislodged from local (and global) energy minima. The Boltzmann distribution of the network state is indeed observed in Monte Carlo trials with the network of Figure B.1,to an accuracy consistent with sample size. Figure B.4 shows the results of one such test in which 999 observations of **x** were recorded at random

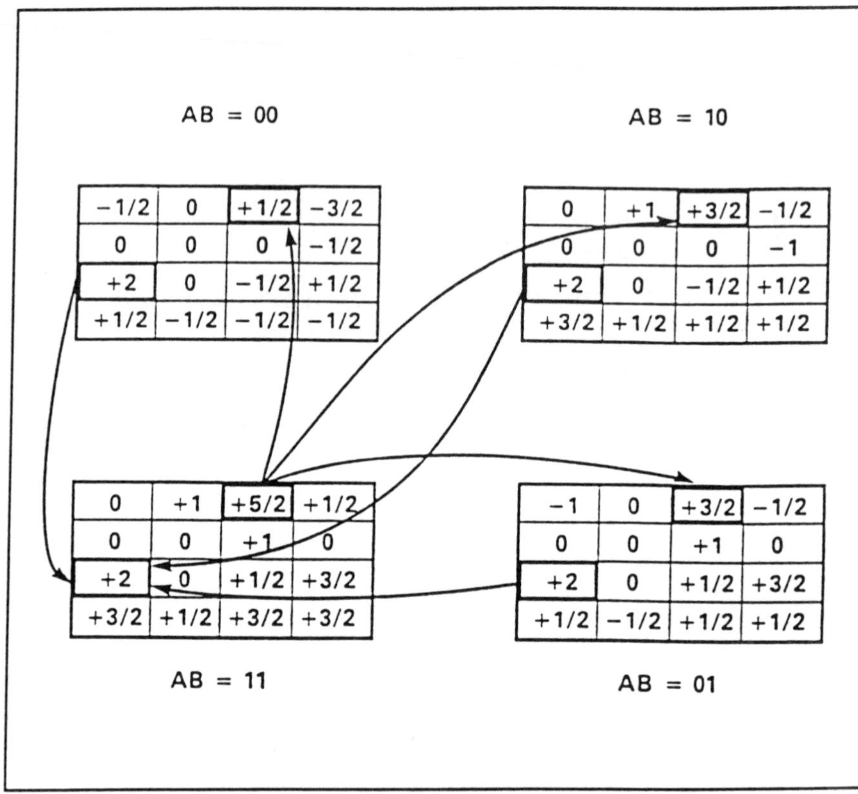

Figure B.3 The negative energy landscapes for the four input conditions have the same format as Fig. 2; but the squares are labeled with -1 times the computational energies.

intervals in the course of ten thousand iterations of the algorithm described above. Here the input is AB = 00 so that the modal probability (i.e., the probability of the most likely state) is P_{12} = Pr(m[**x**] = 12). This test used logistic noise with temperature T=1.

When the noise is not logistic, deviations from the Boltzmann distribution are apparent, especially at lower temperatures. Figure B.5 shows the variation of the modal probability with temperature for each of three noise distributions.

Analysis and Conclusions

One measure of the disparity of two discrete probability densities, **p** and **q**, is the directed divergence (Kullback), or cross-entropy of information for discrimination against **p** in favor of **q**:

$$I(q,p) = \sum_m q_m \ln \left[\frac{q_m}{p_m} \right] .$$

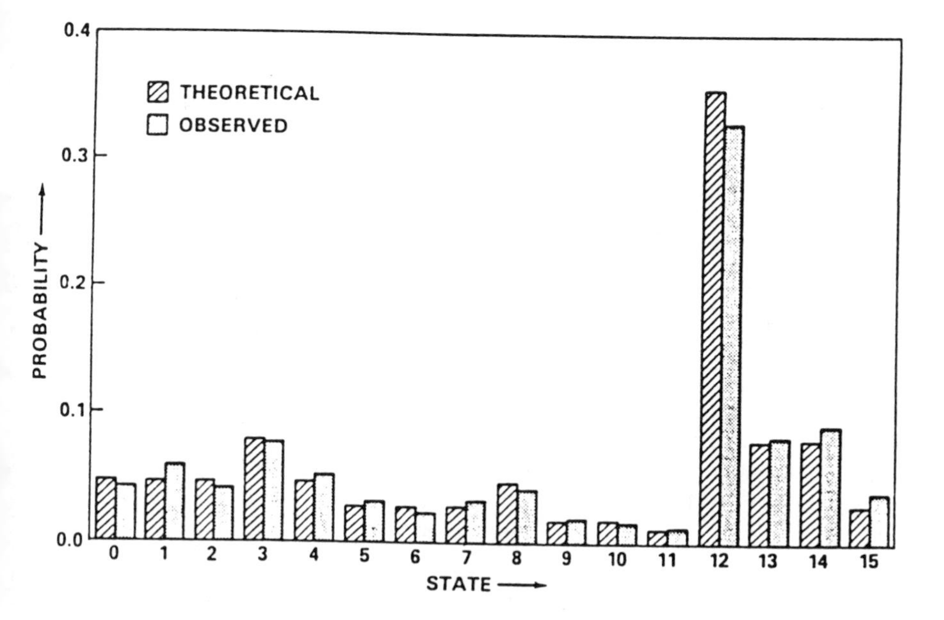

Figure B.4 Distributions of the network state with logistic noise at temperature T=1. Sample distribution is based on 999 observations.

It is well known that, if **q** is a sample distribution, obtained from J independent observations of a random variable with discrete density **p**, $p_m > 0$ for all m $\in$ (0,...M-1), then the product J*I(**q**,**p**) is, in the limit as M/J $\rightarrow$ 0, chi-square with

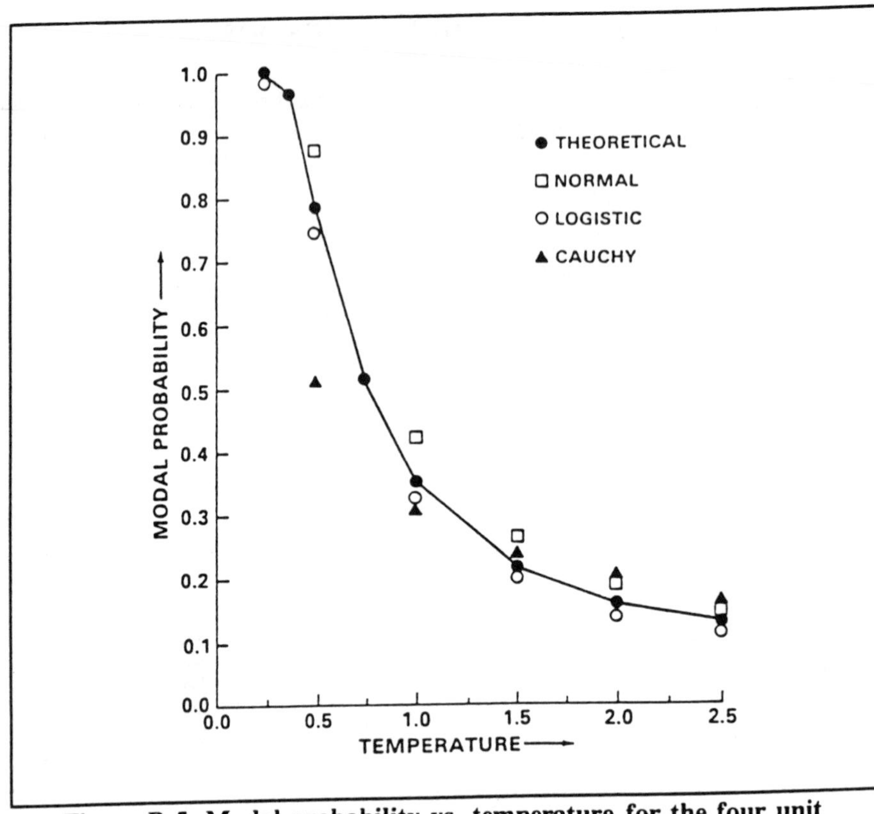

Figure B.5 Modal probability vs. temperature for the four unit "NAND gate" with zero input using three kinds of noise.

M-1 degrees of freedom. Then the mean value of the product J*I is approximately M-1 for large J; and values of J*I (**p**,**q**) in obvious excess of M-1 will tend to refute the null hypothesis **p**.

Table B.2 shows the product of J*I (**p**,**q**) of the sample size and the discrimination information with the Boltzmann distribution as the null hypothesis. Each point represents about a thousand observations of the state of the four-unit "NAND gate" at random intervals in the course of runs of length 10,000. Three different noise distributions are considered with the input AB = 00 at each of five temperatures. The expected value of the cross-entropy statistic is M - 1 = 15 if the null hypothesis pertains. With logistic noise, the observations are below this criterion value in every case. With Cauchy or with

Table B.2 Divergence of the N-sample distribution from the theoretical distribution of the states of the four-unit NAND gate.

Input = (0.0)

T	LOGIS	CAUCHY	NORMAL	(1.1) NORMAL
2.5	13.3	18.9	8.7	18.9
2.0	9.5	15.6	18.6	21.8
1.0	10.1	9.7	24.0	13.7
1.0	6.8	11.1	29.2	20.9
0.5	9.5	206.1	68.7	64.8

ıormal noise, the null hypothesis is clearly rejected at T = 1/2. The case AB = 11 is considered in the right-most column normal noise; and again the est statistic warrants rejection at T = 1/2. These results might be summarized y saying that the asymptotic temperatures, calculated above for nonlogistic ıoises, are reasonable approximations when they equal or exceed unit value.

Appendix C: An Alternative Derivation of the Hinton-Sejnowski Formula for the Weights of a Neural Network Associative Memory

Introduction

The opening paragraphs of the general theory proposed by van Hemmen et al. (1988) will introduce these remarks, which concern networks related to the Hopfield model (Hopfield 1982). These networks are *nonlinear* in the usual sense, consisting of binary threshold elements (Grossberg 1988). They are also nonlinear in the sense defined by van Hemmen et al.: The synaptic kernel is a *nonlinear* function of the vectors that describe the impression of the stored patterns on *pairs* of abstract neurons. Since its introduction by Hopfield, the linear kernel has been studied almost to the exclusion of any other. The desirability of clipped synapses (for the fabrication of analog integrated circuits) clearly motivates one kind of nonlinear prescription. The present discussion, however, is inspired by the statistical physics of the network. It resembles a formula in Rumelhart et al., (1987) attributed to Hinton and Sejnowski (1983), who used evidential reasoning as the setting, and presented a Bayesian statistical argument. Like Hopfield's prescription, which was based on analogy with linear correlation matrix models of associative memory (Hinton and Anderson 1981), the nonlinear prescription is not necessary, but only sufficient for pattern storage. Sufficiency will not be proven here in any rigorous sense; and the question of capacity will remain unanswered. Yet some rather obvious considerations will show that the nonlinear prescription is not radically different from Hopfield's.

The Energy Landscape

As the neurons are spin units, N in number, the state space **S** consists of spin configurations $\{\boldsymbol{\sigma}_a;\ a=1,\ldots,2^N\}$, where $\sigma_a = (\sigma_{1a},\ldots,\sigma_{Na})$, $\sigma_{ia} \in \{-1,1\}$. Let $\mathbf{x} = \{x_1,\ldots,x_N\} \in \mathbf{S}$ be an unindexed spin configuration. At zero temperature, the Glauber dynamics propels the network state downhill in the energy landscape, toward local minima of:

$$H(x) = -\sum_{i<j} J_{ij}\, x_i x_j - \sum_i K_i\, x_i \quad , \tag{1}$$

where external forces make a time-invariant contribution K_i to the local field of the ith unit. The notation $\Sigma_{i<j}$ indicates a sum over all ½N(N-1) distinct pairs. (See chapter 3.)

The Canonical Ensemble

At a finite reciprocal temperature β, the Glauber dynamics gives rise to a homogeneous, irreducible, aperiodic Markov chain on **S**. The dynamical equation of the individual system is written explicitly by Crisanti and Sompolinsky (1988). In a sufficiently large ensemble of such systems, the spin configurations occur with relative frequencies (see Glauber 1963 and van Enter and van Hemmen 1984):[1]

$$P_{1...N}(\boldsymbol{x}) = \exp[-\beta H_{1...N}(\boldsymbol{x})] \,/\, Z_{1...N} \quad . \tag{2}$$

The denominator on the right is the partition function of this canonical ensemble; and the Gibbs distribution that describes the ensemble is also the stationary distribution of the Markov chain.

The Hopfield Model

To create fixed points (the stable states) in **S** at coordinates that match patterns in $\{\sigma_a,\ a=1,\ldots,q\}$, $q << 2^N$, one can form the interaction matrix **J** by setting:

$$J_{ij} = (1/q)\sum_{a=1}^{1} \sigma_{ia}\sigma_{ja}, \quad 1 \neq j \quad ,$$

and $J_{ii}=0$. Ascribing an indicator variable to each spin by $v_{ia} = (\sigma_{ia}+1)/2$,

$$J_{ij} = (1/q)\sum_{a=1}^{q} [2v_{ia}-1)(2v_{ja}-1)] \quad .$$

and $J_{ii} = 0$:

---

[1] $H_{1\ \ldots\ N}$ is the potential of exactly N interacting neurons. If all neurons are interacting, we sometimes write simply H.

Let a pattern correlation matrix **c** be defined by:

$$c_{ij} = (1/q)\sum_{a=1}^{q} v_{ia}v_{ja} \quad .$$

as in chapter 8. Then the prescription is the same as:

$$J_{ij} = 4c_{ij} - 2c_i - 2c_j + 1 \quad , \tag{3}$$

with the shorthand notation $c_i = c_{ii}$. In other words, c_i is the average value of v_i, and c_{ij} is the average of $v_i v_j$, in the pattern ensemble. The prescription is *local*, since the interaction of i and j depends only upon the corresponding components of the pattern correlation matrix. If the pattern set can be decomposed into q/2 spin configuration pairs, the vector sum of each pair being zero, then $c_i = 1/2$ for each i and (4) reduces to:

$$J_{ij} = 4c_{ij} - 1 \quad . \tag{3a}$$

This amounts to storing patterns together with their opposites, a procedure that is perhaps encouraged by the assumption of no external forces, since $H_{1...N}(\mathbf{x}) = H_{1...N}(-\mathbf{x})$ in their absence.

A Nonlinear Kernel

In the notation of (2), $P_{ij}(s_i, s_j)$ is the joint density of the ith and jth spins in a system that is one of N(N-1)/2 systems obtained by regarding N spins as pairwise independent. With reference to (1), define indicator variables almost as before:

$$v_i = (s_i + 1)/2 \quad ,$$

and let P_{ij} be the average value of $v_i v_j$, with $P_i = P_{ii}$. The coupling coefficient is:

$$J_{ij} = \frac{1}{4\beta} \ln \frac{P_{ij}\,(1+P_{ij}-P_i-P_j)}{(P_i-P_{ij})\,(P_j-P_{ij})} \qquad \textbf{(4)}$$

and the forces are:

$$K_{ij} = \frac{1}{4\beta} \ln \frac{P_{ij}\,(P_i-P_{ij})}{(P_j-P_{ij})\,(1+P_{ij}-P_i-P_j)} \qquad \textbf{(5)}$$

by direct calculation. In this way, the three parameters (J_{ij}, K_i, and K_j) can be adjusted to produce an arbitrary joint density of the two spins.

For larger N, one cannot assign 2^N-1 individual state probabilities when only (N+1)N/2 parameters (or degrees of freedom) are present in **J** and **K**. On the other hand, the reduction of the pattern set to a pairwise correlation matrix **C** with (N+1)N/2 free elements gives a one-to-one correspondence that can be exploited in the expectation of collective behavior. This is accomplished by setting **P** = **C**, then constructing **J**(p) in accordance with (4). Thus N(N-1)/2 spin pair systems are created, each of which copies the corresponding local statistics of the pattern ensemble. This replication of the local pattern statistics occurs at a design temperature $1/\beta$ which is the same for all pairs. Hence the nonlinear kernel is:

$$J_{ij} = (1/4) \ln \{ [(c_{ij}(1+c_{ij}-c_i-c_j)]/[(c_i-c_{ij})(c_j-c_{ij})] \} \qquad \textbf{(6)}$$

$i \neq j$, where the temperature is unity in the prescription, since any constant multiple of (6) will give the same fixed points and basins of attraction.

If the patterns are paired, the same assumptions that lead from (3) to (3a) reduce (6) to:

$$J_{ij} = -(1/2) \ln [1/(2c_{ij}) - 1] \ . \qquad \textbf{(6a)}$$

Heuristic Treatment of the External Forces

Equation (5) specifies N(N-1)/2 force pairs which, for any given i, are obviously bound to conflict when the diagonal elements of **c** are not identically 1/2. The Hinton-Sejnowski formulas developed for binary 0-1 (McCulloch-Pitts) neurons, might be adapted to the spin system. The first of these is equivalent to (6). The second prescribes fixed forces:

$$K_i = ln[c_i/(1-c_i)]$$

to bias the units in one direction or the other. The last expression is consistent with (5), except for a constant factor, if each $c_{ij} = c_i c_j$, a condition that means the patterns are uncorrelated. The effect of the forces K_i is the same as that of attributing internal thresholds $-K_i$ to the units. The stability of a configuration **x** will no longer assure that **-x** is stable. This is natural enough in networks of McCulloch-Pitts neurons, where the Glauber dynamics gives rise to expressions formally identical to (1) and (2), but with indicator variables instead of spins, so the energy is not reflection-invariant (Coughlin and Baran 1988). In the following discussion, let each $K_i = 0$, and consider the consequences of using the prescription (6) in the same way that (3) has been employed, without forces or thresholds to bias the spins.

Weakly Correlated Patterns

Let each of q patterns σ_a be generated by N independent Bernoulli trials with success probability 1/2. Then the numbers qc_i are binomial (q,1/2) and the qc_{ij} are binomial (q,1/4). Weakly correlated patterns are thus obtained when the configurations in **S** are selected for storage equiprobably, and q is sufficiently large. Therefore suppose that $c_{ij} = 1/4 + e_{ij}$, $|e_{ij}| << 1$, and $c_i = 1/2 + d_i$ where $|d_i|$ is, likewise, assumed small. The first-order expansion of the logarithm in (6) gives:

$$J_{ij} = 4e_{ij} - 2d_i - 2d_j$$

for each pair. For the Hopfield model (3), the result is the same. Thus the linear and nonlinear prescriptions are approximately equivalent for weakly correlated patterns.

Strongly Correlated Patterns

The prescription (6) cannot be computed when any of the four terms in the argument is zero. The vanishing of c_{ij} means that spins i and j are never both positive in the pattern set. If $1 + c_{ij} - c_i - c_j = 0$, they are never both negative. The vanishing of $c_i - c_{ij}$ means that j always aligns itself with i in the pattern set. Apparently the replication of these pattern statistics cannot be accomplished at a nonzero design temperature with any finite forces and interactions.

Let the stored patterns be paired so that (6a) applies. If **c** contains some off-diagonal elements that lie within a small distance e_{ij} of the endpoints of the interval (0,1/2), the corresponding elements of **J** will be $\pm\frac{1}{2}\log\left(\frac{1}{2e_{ij}} - 1\right)$, which diverges as $e_{ij} \to 0$. This is in contrast to (3a), which limits the strength of the interaction at $c_{ij} = \pm 1$ (the end points of the interval appropriate to (3a)). If spins i and j are always parallel, or always antiparallel, in the pattern set, nonlinear prescription would have to be modified, unless either i or j is deleted. If the elimination of redundant units is impractical for any reason, it would be necessary to clip the nonlinear kernel at arbitrary levels $\pm\log(\eta)$, for some small $\eta > 0$.

Some Illustrations

An example illustrates the deletion of redundant units. Let there be five units and three pattern pairs:

$$\begin{bmatrix} & \sigma_1 & \\ \sigma_5 & & \sigma_2 \\ \sigma_4 & & \sigma_3 \end{bmatrix} \in \left\{ \pm\begin{bmatrix} & + & \\ - & & - \\ + & & + \end{bmatrix}, \pm\begin{bmatrix} & - & \\ + & & - \\ - & & + \end{bmatrix}, \pm\begin{bmatrix} & + & \\ + & & + \\ - & & + \end{bmatrix} \right\}$$

Spins four and five are always antiparallel; so delete the fifth and obtain the reduced problem:

$$\begin{bmatrix} \sigma_1 & \sigma_2 \\ \sigma_4 & \sigma_3 \end{bmatrix} \in \left\{ \pm\begin{bmatrix} + & - \\ + & + \end{bmatrix}, \pm\begin{bmatrix} - & - \\ - & + \end{bmatrix}, \pm\begin{bmatrix} + & + \\ - & + \end{bmatrix} \right\}$$

in which all patterns feature a single triangle of parallel spins; but the right lower triangle is not a desired pattern. The 4×4 pattern correlation matrix **c** is 1/6 times:

$$\begin{bmatrix} 3 & 2 & 2 & 2 \\ 2 & 3 & 1 & 1 \\ 2 & 1 & 3 & 1 \\ 2 & 1 & 1 & 3 \end{bmatrix}$$

Using (6a), the interactions are:

$$2J_{ij} = \begin{cases} ln2 & if\ i{=}1\ and\ j{>}i \\ -ln2 & if\ i{>}1\ j{>}i \end{cases} .$$

Thus the energy is -½ ln2 for each of the six desired patterns, and +3/2 ln2 fc the undesired case in which the last three spins are in alignment. Using (3a the interactions are the same but for a common factor. The desired patterns ar the only stable states in these networks.

Another example illustrates how the linear and nonlinear kernels ca give essentially the same result. Let N=9 and make six pattern pairs, namely

```
+ + +     - - -     - - -     + - -     - + -     - - +
- - -     + + +     - - -     + - -     - + -     - - +
- - -     - - -     + + +     + - -     - + -     - - +
```

and their opposites. The nondiagonal elements of c are either 1/3 (if i and share the same row or column) or 1/6 (if not). Thus the Hopfield model lead to J_{ij} which are either +1/3 (if same row or column) or -1/3 (if not). Th alternative (6a) leads to $2J_{ij}$ being +ln2 (if same) or -ln2 (if not). The energ landscapes are identical, except for a scale factor that affects neither th coordinates of the fixed points nor their basins of attraction. The stable state of these networks include all the desired patterns and superpositions of patter pairs, like:

```
+ + +        - + -
+ - -  ,     + + +  ,    etc.
+ - -        - + -
```

Now let the pattern set include only those six configurations jus displayed (not their opposites). The diagonal elements of c are all 1/3; and th off-diagonal elements are either 1/6 (if same row or column) or zero. thus th linear kernel (3) is exactly as before; but the nonlinear kernel, clipped at ln(η) is:

$$J_{ij} = \begin{cases} (ln3)/2 & if\ same \\ -|\ln\eta| & if\ not. \end{cases}$$

The "strong inhibition" thus produced does not change the fixed points in thi case.

APPENDIX D: Optimization in Cascaded Boltzmann Machines with a Temperature Gradient: An Alternative to Simulated Annealing

Introduction

The application of neural networks to constrained optimization has fallen out of favor in recent years. On the one hand, investigators using the continuous, deterministic dynamics of symmetrically interconnected "neural circuits" have reported difficulty in obtaining the "good (suboptimal) solutions" found by Hopfield and Tank (1985, 1986). On the other, simulated annealing in "Boltzmann machines" made of binary threshold units with stochastic dynamics has been problematic since Geman and Geman (1984) guaranteed convergence to global optima with an impractically slow cooling schedule. Subsequent work on the computational thermodynamics of symmetric neural networks has tended toward the formulation of expedient cooling schedules (Aarts and Korst 1989).

We offer a radical variation of the stochastic model in which the slow cooling of the single network over time is replaced by a time-invariant temperature gradient across cascaded subnets. Each subnet by itself is an isothermal Boltzmann machine with binary units and reciprocal connections. These subnets are series-coupled with one-way retinotopic connections. The simplest case features two identical subnets, the first being at temperature $T > 0$ and the second at zero temperature. Each unit in the warmer subnet influences its counterpart in the cold subnet through a one-way connection (or spanning link) that transmits a bias proportional to the sign of the first. As the warmer subnet wanders ergodically through its state space, it attracts the cold subnet, causing the energy of the latter to dwell on successively lower plateaus. The candidate solutions registered by the cold subnet evolve toward global optima in a few hundred sweeps of the network. The method uses only local information and is well suited to concurrent execution on two or more asynchronous machines.

Background

Hinton and Sejnowski (1983) invented the Boltzmann machine b
substituting stochastic neurons for the deterministic units of the original Hopfiel
model. As was shown in chapter 3, every state of a symmetricall
interconnected network of 0-1 binary units has a computational energy and th
asynchronous operation of the individual units gives rise to collective behavio
as the network state moves downhill in energy — eventually reaching fixe
points (stable states) at local energy minima. The original Hopfield model i
usually realized in discrete time simulations by selecting units in a random seria
order and updating them according to the decision rule of the McCulloch-Pitt
neuron. For instance, if the ith unit out of N is selected for update at time t, it
state (activation) at t+1 is:

$$x_i(t+1) = \begin{cases} 1 \ \textit{if} \sum_{j=1}^{N} W_{ij}\, x_j + U_i > 0 \\ \\ 0 \ \textit{otherwise} \end{cases} \tag{1}$$

where **W** is the symmetric weight matrix, $W_{ii} = 0$, and **U** is a vector of biases
(A bias of U_i is equivalent to a threshold of $-U_i$.) The energy is:

$$H(t) = H[\mathbf{x}(t)] = -\sum_{i<j} W_{ij} x_i(t) x_j(t) - \sum_i x_i(t)\, U_i \tag{2}$$

in which the first sum on the right side is over all (unordered) pairs of units
Hopfield (1984) put aside this stochastic model after showing that simila
behavior could be obtained in "neural circuits" in which nonlinear amplifier
feed back to each other through a symmetric array of coupling conductances (th
weights). In the high-gain limit, where the amplifiers become hard limiters, th
stable states of the electronic circuit are the same as those of the correspondin
stochastic model, since the energy (Lyapunov) functions are the same. Hopfiel
and Tank (1985, 1986) then exhibited neural circuits for solving optimizatio
problems. In these optimization circuits the stable states represent candidat
solutions when they satisfy certain constraints. For example, in the n-cit
traveling salesman problem (TSP), the circuit features a square n-by-n array o
units; and every state represented by a permutation matrix corresponds to a vali
tour. Kahng (1989), following Wilson and Pawley (1988), tried to confirm th
preference for "good (suboptimal) solutions" reported by Hopfield and Tank
Running ten-city TSPs with the canonical weight prescription, Kahng found th
"tour quality is good when a valid tour is found, but unfortunately only abou
15% of the outputs are valid tours."

Other investigators have returned to the stochastic model and added a new dimension of randomness to the dynamics. By disclosing a learning algorithm for "Boltzmann machines", Ackley, Hinton and Sejnowski (1985) drew attention to an idea sketched earlier by Hinton and Sejnowski (1983), who observed that a symmetrically interconnected network of binary units "suffers from the standard weakness of gradient descent methods: It gets stuck in local [energy] minima that are not globally optimal." In the Boltzmann machine, the decision rule (1) is altered, for the case $\Delta H > 0$:

$$x_i(t+1) = \begin{cases} 1 \textit{ with probability } P[\Delta H_i(t)] \\ 0 \textit{ with probability } 1 - P[\Delta H_i(t)] \end{cases} \tag{3}$$

where:

$$P(y) = \frac{1}{1 + e^{-\beta y}} \tag{4}$$

and:

$$\Delta H_i = \sum_{j=1}^{N} W_{ij} x_j + U_i$$

is the energy gap associated with the toggling of the ith unit. (See chapter 3.) The parameter $T = 1/\beta$ acts like temperature and the ergodic wandering of the global state $\mathbf{x}(t)$ leads in the long run to a stationary Boltzmann distribution, as:

$$\lim_{t \to \infty} Pr(X(t) = x) = Z^{-1} \exp(-H[x]/T) \tag{5}$$

where, as in chapter 7, $H(\mathbf{x})$ is the energy of state $\mathbf{x}$ and the partition function Z is the sum over all 2N states. (This normalizes the discrete density.) Equation (3) reduces to (1) in the special case $T = 0$.

The computational thermodynamics of symmetrically interconnected neural optimization networks usually relies on equation (5) which, strictly speaking, applies if and only if T is a strictly positive constant. The isothermal Boltzmann machine visits a given state with a relative frequency that depends on the energy of the state. At $T = 0$, the probability given by (5) vanishes except for the one state that occupies a global energy minimum (or a collection of states that share the same distinction). The objective then is to cool the network so slowly that it never departs too drastically from the thermal equilibrium described by (5).

Simulated Annealing in Boltzmann Machines

Simulated annealing is a way to find the global extremum of a function that has many local extrema and may not be smooth (Bohachevsky et al. 1986). The method is a biased random walk that samples the objective function in the space of the independent variables. Simulated annealing with "fast computing machines" traces its roots to the work of Metropolis et al. (1953). It can be shown that the Boltzmann machine implements the Metropolis algorithm, equation (3) describing a random walk with acceptance probabilities determined with reference to the energy function (2).

Geman and Geman (1984) proposed an annealing schedule of the form $T(t) = T_0/\ln(1 + t)$, $t = 1, 2, \ldots$, and showed that it guaranteed convergence to global energy minima. Unfortunately, the convergence time with this cooling schedule is on the order of 2N sweeps of the network, a sweep being defined as N consecutive iterations of equation (3). Thus the annealing process takes about as long as an exhaustive state space search. Aarts and Korst (1989) have provided theoretical motivation for more expedient cooling schedules, like:

$$T(s) = T_o(1 - q)^s, \quad 0 < q << 1 , \tag{6}$$

in which $s = 0, 1, 2, \ldots$ can count the sweeps or the time steps. They note that good (suboptimal) performance is typical of their computational experience.

We used n-to-n task assignment problems of the kind suggested by Hopfield and Tank (1986b) to test and evaluate the geometric cooling schedule (6). The problem is to assign n tasks to n workers in a way that maximizes total productivity, when the task-specific production rates are given for each worker. Let m index the tasks and let k index the workers. A real number $U_{k,m}$, generally nonnegative, gives the productivity of the kth worker on the mth task. The problem is to assign one $m^*(k)$ to each k in a way that maximizes the aggregate productivity:

$$V = \sum_{k=1}^{n} U_{k,\, m^*(k)} \tag{7}$$

The neural solution involves an n x n array of n^2 = N units which are usually identified by row (worker number) and column (task number). These units can be indexed serially in any manner so long as the weight matrix, **W**, is given by:

$$W_{ij} = \begin{cases} -C \text{ if } row(i) = row(j) \text{ or if } col(i) = col(j), \quad i \neq j, \\ 0 \text{ otherwise} \end{cases} \tag{8}$$

U_i is the production rate (indexed serially) of the worker indexed by k = row(i) given the task indexed by m = column(i). When:

$$C > \max_{1 \le i \le N} U_i = U_{max} \tag{9}$$

the strong inhibition (8) guarantees that only the permutations of the n x n identity matrix are the stable states of the X-array. The energy of stable state vector **x** is simply the inner product -U•**x**, which is the negative of the aggregate production rate.

We wrote a QuickBASIC program called TASK PLANNER for AT-compatible computers with color (E/VGA) graphics. TASK PLANNER includes an editor for creating and viewing Task Assignment Charts in which the elements of the U-array are nonnegative with up to three digits of resolution. These charts are saved as *.TAC files which provide input to the neural optimization routine. TASK PLANNER finds the largest task-specific production rate on the chart and divides all the other elements by this amount so that Umax = 1. Then the strength of the inhibitory links is set at C = 1.2 consistent with (9). TASK PLANNER presently handles U-arrays of dimension up to 18-by-16.

TASK PLANNER uses the geometric cooling schedule (6) with initial temperature T_0 = 0.70 and adjustable rate parameter q = ALPHA. The user inputs a nonnegative integer r and the program sets:

$$q(r) = \frac{0.25}{1 + 2r} \tag{10}$$

Thus the cooling rate increases with r = 0, 1, 2, ... Beginning in a random initial state, the network is swept once at T = T_0, once at T = $T_0(1 - q)$, and so on, up to:

$$S(q) = integer\frac{2.4}{q} + 10 \quad sweeps \tag{11}$$

by which time convergence is almost guaranteed. The assignments are then displayed and the figure of merit (7) is computed.

We constructed a 10x10 Task Assignment Chart in which the 100 entries were independent random numbers, uniformly distributed on the integers from 0 to 999. One might try to solve the problem by taking the tasks in any sequence and assigning them the best-suited worker who has not been previously assigned. It can be shown by elementary statistical methods that the expected value of the figure of merit (7) attained by this "greedy" algorithm is approximately:

$$E(V_{greedy}) = n(n-1)/(n+1)$$

when all the U_i are i.i.d. U(0, 1). The greedy solution to the given problem might then have aggregate productivity of about 818 (after substituting n = 10 in the last equation and scaling up by a factor of 100). TASK PLANNER was used to find six solutions at each value of the cooling rate parameter ALPHA = q(r) as r was varied from 0 to 9 in equation (10). Figure D.1 shows the averages of the six results at each ALPHA. The best solution, V = 847, was obtained on the first trial with r = 9. The first trial with r = 0 yielded V = 812, very close to the mean for the greedy algorithm; but all subsequent trials with this very fast cooling schedule gave poorer results. Figure D.1 suggests that it is necessary to use ALPHA > 0.016, and to cool over more than S(0.016) = 160 sweeps, in order to get results which, on the average, are competitive with the greedy algorithm.

Optimization in Cascaded Boltzmann Machines with a Temperature Gradient

We offer a radical variation of the stochastic model in which the slow cooling of the single network over time is replaced by a time-invariant temperature gradient across cascaded subnets. Each subnet by itself is an isothermal Boltzmann machine with binary units and reciprocal connections. These subnets are series-coupled with one-way retinotopic connections. (By "retinotopic" we mean that a spanning link runs from unit i in subnet k to unit i in subnet k+1 and so on.) Let:

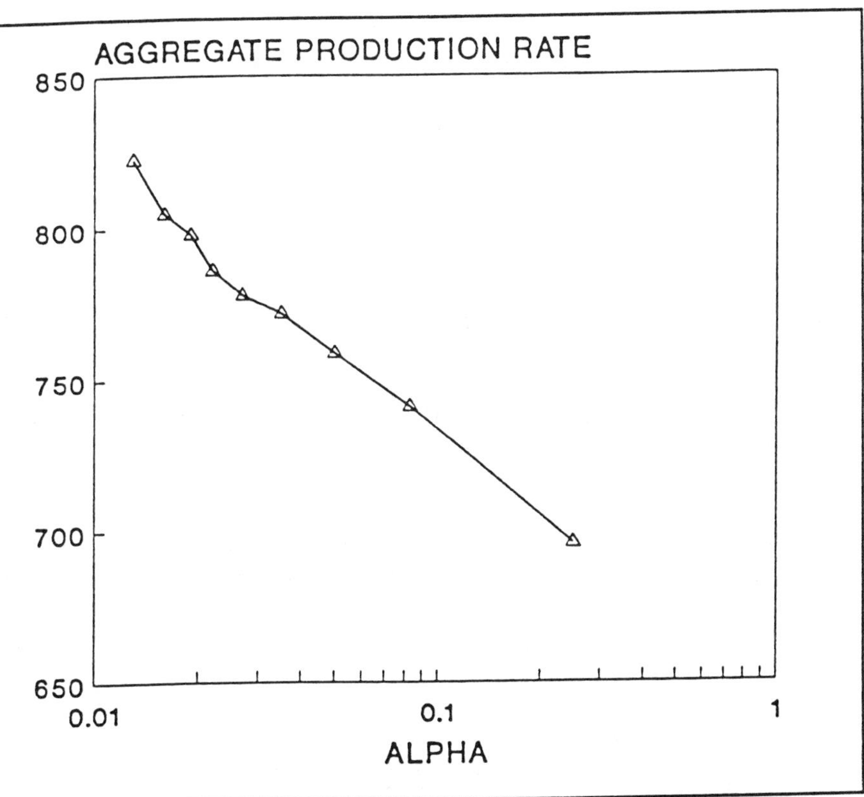

Figure D.1 Mean solution quality versus cooling rate parameter (ALPHA) using TASK PLANNER to solve a 10-to-10 assignment problem with random individual task-specific rates.

$x_{i,k}(t)$ = the binary 0-1 state of the ith unit in the kth subnet at time t.

The dynamical equations for our system are:

$$C_{i,k} = \sum_{j=1}^{N} W_{ij}\, x_{j,k}(t) - U_i + T_k \ln\left(\frac{1}{R_k(t)} - 1\right) + B_{k-1}\,[2x_{i,k-1}(t^*_{k-1}) - 1]$$

$$x_{i,k}(t+1) = \begin{cases} 1 & \text{if } C_{ik} > 0 \\ 0 & \text{otherwise} \end{cases} \qquad (12)$$

$i,j \in \{1, 2, ..., N\}$, $k \in \{1, 2, ..., K\}$, where T_k = the temperature of the kth subnet, $R_k(t)$ = a computer-generated random number, uniform on the unit interval and independent for each k and t, and B_{k-1} = the strength of the spanning links from subnet k-1 to subnet k, which may depend on temperature. ($B_0 = 0$.)

$$B_1 = (1 - T)b \tag{13}$$

Note that if $X \in \{0,1\}$, then 2x-1 is the sign of x. The transformation T ln(1/R - 1) produces a logistic random variable with distribution function (4) (see page D-3) from the computer-generated R. Its addition to the input of the binary unit is equivalent to invoking the test (3) in the manner of the stochastic unit (Baran, 1988). Imposition of the temperature gradient:

$$T_1 > T_2 > ... > T_K$$

then amounts to injecting white noise, with the logistic distribution (4), into all the units, making the noise intensity decline as the subnet index k increases. This scheme is pictured in Figure D.2 where, borrowing from the terminology of Metropolis et al., the temperature gradient is maintained by separate "heat baths" for the cascaded subnets.

TWONETS

We wrote another BASIC program, called TWONETS, to experiment with the simplest case of equation (12), which is K = 2 identical subnets. Like the TASK PLANNER described above, TWONETS finds solutions to n-to-n assignment problems. Thus the weight matrix W is again given by equation (8) and the bias vector **U** consists of i.i.d. U(0, 1) random variables.

TWONETS sweeps the two subnets alternately. In other words, after iterating equation (12) N times for k=1, it then does the same for the cold subnet (k = 2) after updating the array $[x_{i1}(t)]$, which figures in the bias term, at times $t_1^* = N, 2N, 3N$, etc. In principle one could substitute t for t_{k-1}^* in equation (12). By sweeping the subnets alternately we simulated the behavior of a parallel, asynchronous implementation in which the subnets reside in physically separate computers that communicate once every N unit updates.

We took $T_1 = T > 0$ and $T_2 = 0$. Since T = 1 is sufficiently hot to make the first subnet state change with almost every unit update, we let the strength of the spanning links be a linear function of the temperature of the

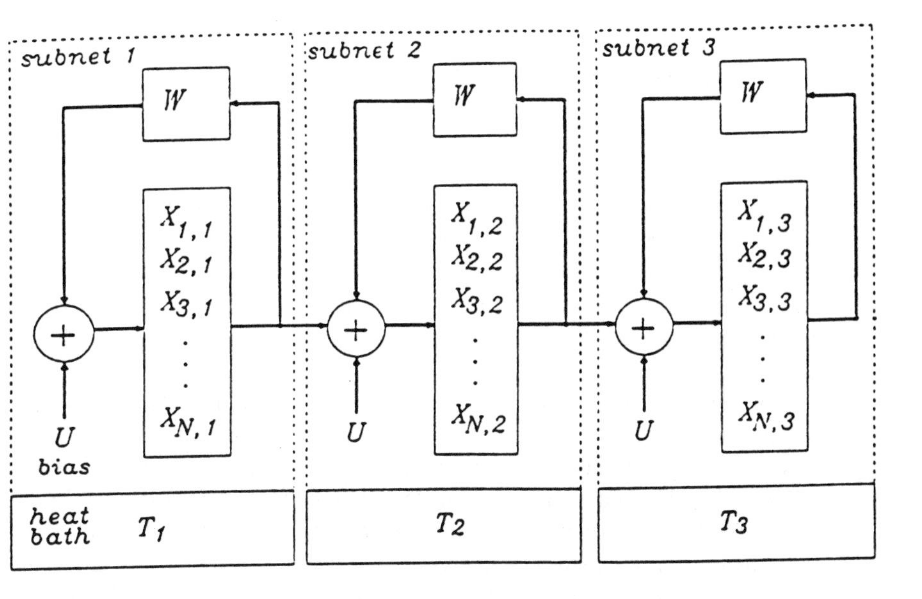

Figure D.2 Block diagram interpretation of equation (12) for the case of K = 3 identical subnets. The temperature gradient is imposed as $T_1 > T_2 > T_3 = 0$.

warmer subnet, namely for a fixed parameter b: $C > b > 0$. The results shown below were obtained with $C = 1.1$ and $b = 0.75$. Larger values of b caused the cold subnet to be too strongly attracted by the warm one, so that it followed the latter uphill in energy. Smaller values of b weaken the strength of the spanning links and allow the cold subnet to ignore the guidance provided by the warm one.

Figure D.3(a) shows the result of a typical experiment with $T = 0.2$. The energy of the warm subnet wanders between -4 and -8 dimensionless units in the course of 150 sweeps. As its energy declines over the first 50 sweeps, it "pulls" the state of the cold subnet toward it. This influence actually causes the energy of the cold subnet to fall below the lowest value attained by the warm one. As the warm subnet wanders off (after $t = 60N$), the cold one continues to recommend the best solution it has reached. This kind of collective behavior is not guaranteed every time. Disappointing performance occurs when the cold subnet leaves more nearly optimal states in pursuit of the warm one, as in Figure D.3(b), and when the cold subnet repeatedly fails to follow the warm one downhill in energy. These difficulties were the exception rather than the rule when the T was in the vicinity of 0.2.

The optimal choice of T, the temperature of the warmer subnet, can be determined by numerical experimentation. Guided by results of which Figures D.3(a) and D.3(b) are representative, the objective is to cause the cold subnet to follow the warm one downhill in energy -- but to remain in deep energy wells, corresponding to good solutions, after the warm subnet wanders off into its state space. With this in mind, the numerical trials need to run long enough for good solutions to be found (and possibly abandoned).

Figure D.4 shows how the energies of the two subnets evolve over 200 sweeps in the course of solving a 10-to-10 assignment problem. Each plotted point is the average of ten independent trials each involving a different U-array. With $T = 0.25$, the cold subnet is typically attracted to (within a few tenths of) $H = -8$ in the first 100 sweeps. Recalling that 8.18 is the expected value of the figure of merit (V) obtained by the greedy algorithm in problems of this type, and that $H[\mathbf{x}] = -V$ when $\mathbf{x}$ is a stable state, it appears that TWONETS requires about 100 sweeps to get good results with the given parameters.

Figure D.5 shows the variation of energy with T. Each plotted point represents the average of ten independent trials each of which extended to $t = 200N$ per subnet. With reference to the preceding figure, we know that 200 sweeps is well into the "saturation region", at least for $T = 0.25$. With reference to equation (13), note that $T=1$ decouples the subnets. Thus the average final energy of the cold subnet following the dynamical equation (1) is about -6. Taking $T = 0.25$ reduces the average energy to about -8. The experiments summarized in this figure were carried out after translating the original TWONETS into FORTRAN for execution on a Micro-VAX.

Conclusion

TWONETS proves the feasibility of doing optimization in cascaded Boltzmann machines with a temperature gradient. Its performance in small, randomly generated n-to-n assignment problems is comparable to that of the usual simulated annealing method with the geometric cooling schedule. Both methods yield solutions that are competitive with the greedy algorithm. Both methods require between 100 and 200 sweeps to attain this performance level. The same conclusions would probably have been reached if we had chosen a different type of optimization problem, like the TSP, to benchmark the two techniques.

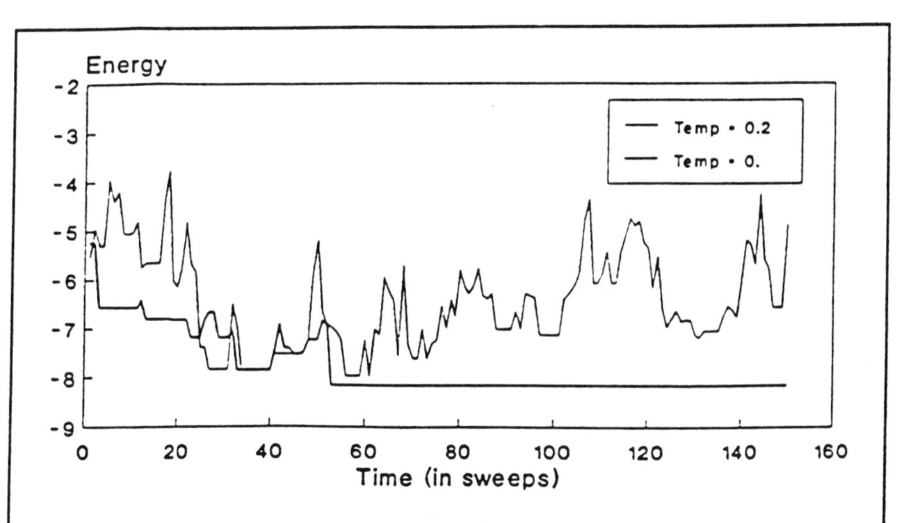

Energy versus time in identical subnets for computing solutions to a 10-to-10 task assignment problem. Subnet 1 at T • 0.2 drives subnet 2 at T • 0 through one-way retinotopic connections.

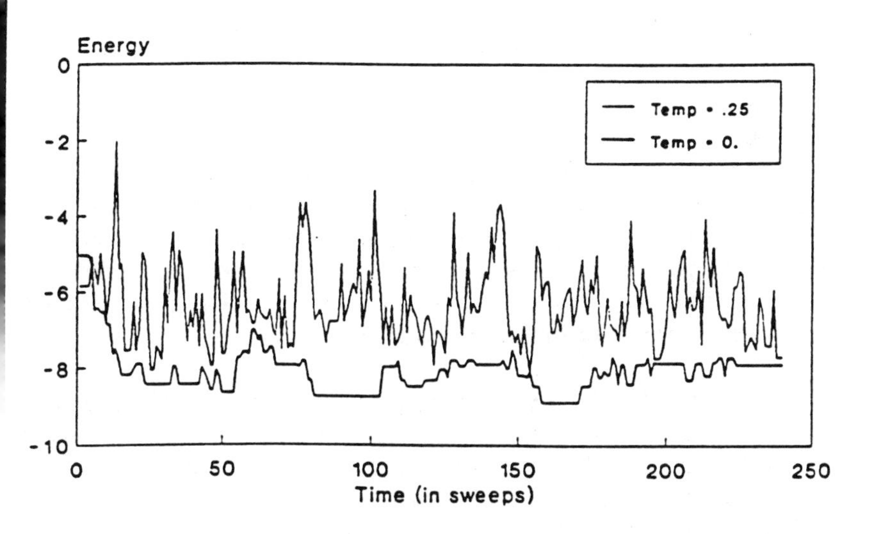

Energy versus time in identical subnets for computing solutions to an 11-to-11 task assignment problem.

Figure D.3 Energy vs. time in identical subnets.

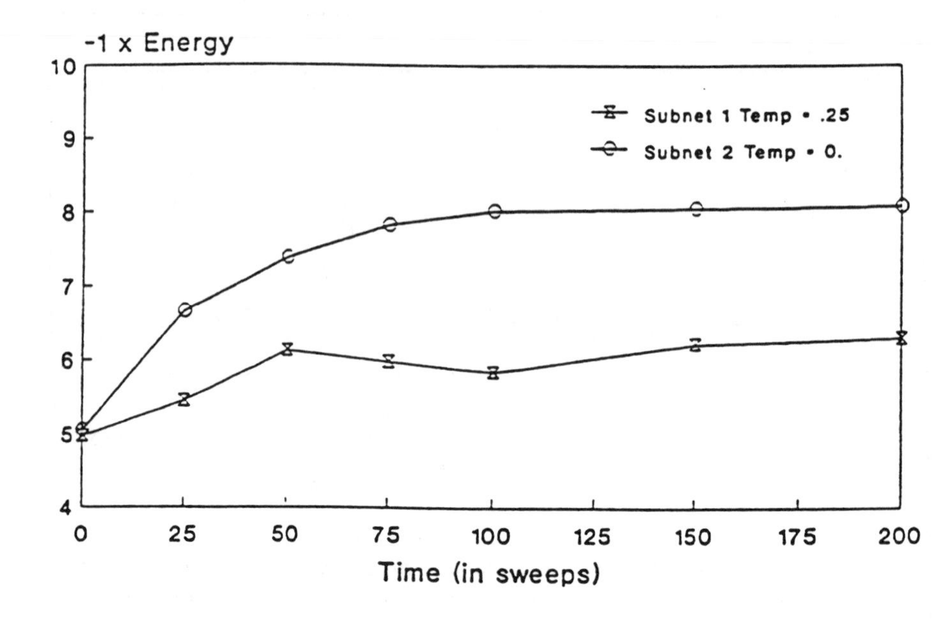

Figure D.4 Negative Energy vs. time in 10x10x2 optimization nets. Each plotted point is the mean of 10 independent trials.

Can our method outperform simulated annealing? The formula (13) for the strengths of the spanning links gave encouraging results; but we can offer no proof of its optimality. In particular, we would like to know if two or more cascaded subnets, perhaps with a better formula than (13), can consistently find better solutions than the greedy algorithm, or converge to solutions of the same quality in fewer sweeps than simulated annealing.

We suspect that the answer is negative in so far as the growth of average solution quality with number of sweeps reflects the state space distance from the random initial state to the solution state. On the other hand, it will not be difficult to construct software models in which more than two subnets are linked. Such models can show whether solution quality at the output subnet increases with the gradualness of the temperature gradient across the subnets driving it.

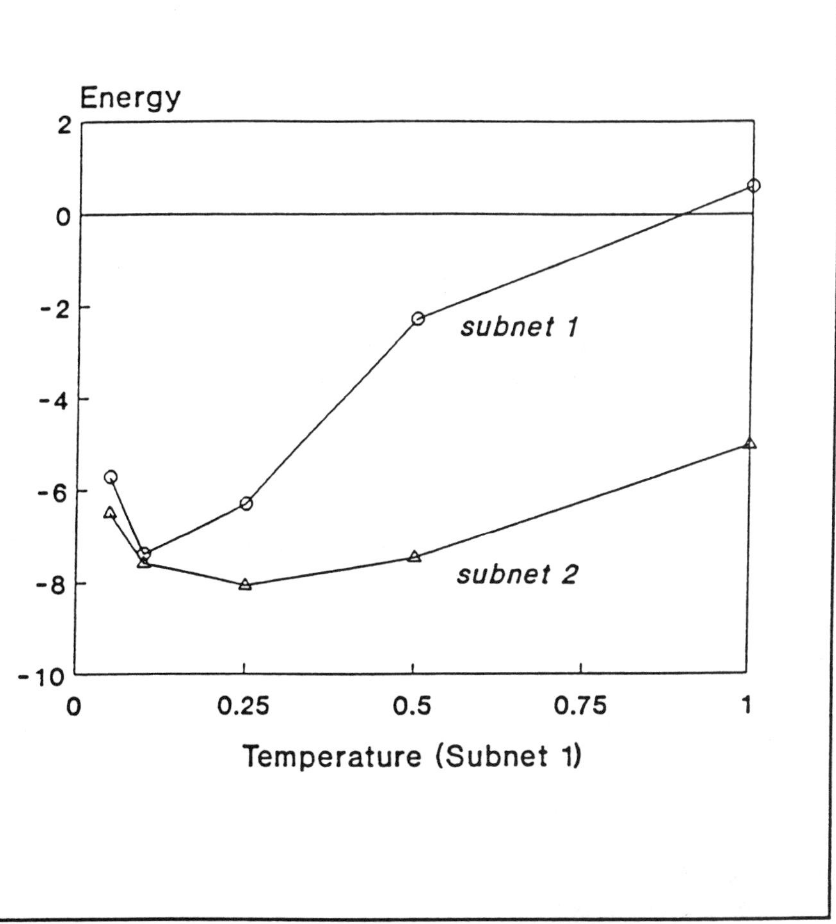

Figure D.5 Energy vs. temperature in 10x10x2 optimization nets. Each plotted point is the mean of 10 independent trials each using a random bias array. Temperature of subnet 2 is 0.

APPENDIX E:

Basic Program for Computing a Sample Network

```
100 REM  This is to compute the stationary distribution in the three-unit
         network described by connectivity matrix
            ( 0  1 -1)
            ( 1  0  1)
            (-1  1  0)
          using McCulloch-Pitts neurons and Glauber
          dynamics.

110 DIM W(3,3)
120 W(1,2)=1
121 W(2,1)=1
122 W(1,3)=-1
123 W(3,1)=-1
124 W(2,3)=1
125 W(3,2)=1
130 RANDOMIZE
140 INPUT LAST,TEMP
141 PRINT ""
150 FOR COUNT=1 TO LAST
160     GEO=INT(-10*LOG(RND))
190     FOR T=1 TO GEO
200       I=INT(3*RND)+1
210       U=TEMP*LOG(1/RND-1)
230       FOR J=1 TO 3
240          U=U+W(I,J)*X(J)
250       NEXT J
260       X(I)=1
270       IF U>0 THEN 290
280          X(I)=0
290     NEXT T
```

```
300      STATE=1+X(1)+2*X(2)+4*X(3)
310      PROB(STATE)=PROB(STATE)+1
330  NEXT COUNT
340  FOR STATE=1 TO 8
350      PROB=.001*INT(1000*PROB(STATE)/COUNT)
360      PRINT STATE-1, PROB
370      SUM=SUM+PROB
380  NEXT STATE
390  PRINT "__________________"
400  PRINT "sum",SUM
410  END
```

```
print 1/log(2)
 1.442695
Ok

RUN
Random number seed (-32768 to 32787)? 255 ?
1000,1.4427

0          8.300001E-02
1          7.900001E-02
2          .102
3          .19
4          9.100001E-02
5          .045
6          .217
7          .185
------------------
sum  .9920001
```

REFERENCES

Aarts, E. and Korst, J. 1989. *Simulated Annealing and Boltzmann Machines.* New York: John Wiley and Sons.

Ackley, D. H., Hinton, G. E., and Sejnowski, T. J. 1985. A learning algorithm for Boltzmann machines. *Cognitive Science 9*: 147-169.

Akiyama, Y., Yamashita, A., Kajiura, M. and Aiso, H. 1989. Combinatorial optimization with Gaussian machines. *Proceedings of the International Joint Conference on Neural Nets.* 1:533-540 [IEEE #89CH2765-6].

Babcock, K. L. and Westervelt, R. M. 1987. Dynamics of simple electronic neural networks. *Physica 28D*:305-16.

Baran, R. H. 1988. Comments on "A new theoretical and algorithmical basis for estimation, identification and control". *Automatica 24* (2): 283-7.

___________. 1989a. A collective computation approach to automatic target recognition. *Proceedings International. Joint Conference on Neural Nets.*, 1:39-44.

___________. 1989b. A neural network approach to data fusion in automatic target recognition. *International Journal of Neural Nets: Research & Applications 1*(2):68-77.

Bellman, R. 1953. *Stability Theory of Differential Equations.* New York: McGraw-Hill.

Blackwell, D. and Girshick, M. A. 1954. *Theory of Games and Statistical Decisions.* New York: John Wiley and Sons.

Bohachevsky, I. O., Johnson, M. E. and Stein, M. L. 1986. Generalized simulated annealing for function optimization. *Technometrics 28*(3):209-17.

Caianiello E. R. 1986. Neuronic equations revisited and completely solved. In *Brain Theory.* Edited by G.Palm and A.Aertsen, 147-160 New York: Springer Verlag

Carr, W. N. and J. P.Mize. 1972. *MOS/VLSI Design and Application*. New York: McGraw-Hill.

Cipra B. 1987. An introduction to the Ising model. *American Mathematical Monthly 94*:937-959.

Cohen, M. A. and Grossberg, S. 1983. Absolute stability of global pattern formation and parallel memory storage by competitive neural networks. *IEEE Trans. Systems Man & Cybernetics 13* (5): 815-25.

Coughlin, J. P. and Baran, R. H. 1988. Remarks on neural networks and spin glass models. *American Mathematical Monthly 95*:631.

Crisanti, A. and Sompolinsky, H. 1988. Dynamics of spin systems with randomly asymmetric bonds: Ising spins and Glauber dynamics. *Physical Review A 37*(12):4865-74.

Dammasch, I. E. and Wagner, G. P. 1984. On the properties of randomly connected McCulloch-Pitts networks. *Cybernetics and Systems 15*:91-117.

Dobrushin, P. L. 1968. The description of a random field by means of conditional probabilities and conditions of its regularity. *Theory of Probability and Its Applications 8*(2):197-224.

Drake, A. W. 1967. *Fundamentals of Applied Probability Theory*. New York: McGraw-Hill.

Farley B. G. 1960. Self-organizing models for learned perception. In *Self-Organizing Systems*. Edited by M.Yovits and S.Cameron New York: Pergamon.

Feller W. 1957. *An Introduction to Probability Theory and Its Applications*. 1, 2nd ed. New York:John Wiley and Son

Geman, S. and Geman, D. 1984. Stochastic relaxation, Gibbs distributions, and the Bayesian restoration of images. *IEEE Transactions in Pattern Analysis & Machine Intelligence*. 6:721-741.

Gibbs, J. W. 1902. *Elementary Principles in Statistical Mechanics*. [Yale University Press, New Haven, Connecticut (reprinted by Dover Publications, New York (1960).

Glauber, R. J. 1963. Time-dependent statistics of the Ising model. *Journal of Mathematical Physics 4*(2):294-307.

Grossberg, S. 1978. Decisions, patterns and oscillations in nonlinear competitive systems with applications to Volterra-Lotka systems. *Journal of Theoretical Biology 73*: 101-30.

Grossberg, S. 1988. Nonlinear neural networks: principles, mechanisms and architecture. *Neural Networks 1*(1):17-61.

Hartman, P. 1964. *Ordinary Differential Equations*. New York:John Wiley and Sons.

Hebb, D. O. 1949. *The Organization of Behavior*. New York:John Wiley and Sons.

Hinton, G. E. and Anderson, J. A. eds. 1981. *Parallel Models of Associative Memory*. Hillsdale, New Jersey:Lawrence Erlbaum Associates.

Hinton, G. E. and Sejnowski, T. J. 1983. Optimal perceptual inference. *Proceedings IEEE Conference on Computer Vision & Pattern Recognition*. 448-53. Los Alamitos, California:IEEE Computer Society Press.

Hirsch, M. W. and Smale, S. 1974. *Differential Equations, Dynamical Systems and Linear Algebra*. San Diego,California: Academic Press.

Hopfield J. J. 1982. Neural networks and physical systems with emergent collective computational abilities. *Proceedings National Academy of Sciences USA 79*:2554-8.

____________ 1984. Neurons with graded responses have collective computational properties like those of two-state neurons. *Proceedings National Academy of Sciences USA 81*:3088-92.

Hopfield, J. J. and Tank, D. W. 1985. "Neural" computation of decisions in optimization problems. *Biological Cybernetics 52*:141-52.

____________________________ 1986a. Simple "neural" optimization networks: an A/D converter, signal decision circuit, and a linear programming circuit. *IEEE Transactions Circuits & Systems 33*(5):533-41.

______________ ______________ 1986b. Collective computation in neuronlike networks. *Scientific American* (Dec. 1987) pp. 104-14.

______________________________ 1986c. Computing with neural circuits: a model. *Science 233*:625-32.

Kahng A. B. 1989. Traveling salesman heuristics and embedding dimension in the Hopfield model. *International Joint Conference on Neural Nets*. Washington D.C.

Kirkpatrick S. et al. 1983. Optimization by simulated annealing. *Science 220*:671-80.

Kohonen T. 1977. *Associative Memory: A System-Theoretical Approach*. New York:Springer Verlag.

Kohonen, T., E. Oja and P. Lehtio.1981. Storage and processing of information in distributed associative memory systems. In *Parallel Models of Associative Memory*. Edited by E.Hinton, & J.A.Anderson 105-43. Hillsdale, New Jersey:Lawrence Erlbaum Associates.

Komlos, J. and R. Paturi. 1988. Convergence results in an associative memory model. *Neural Networks 1*(3):239-50.

Kramers, H. A. and G. H. Wannier. 1941. Statistics of the two dimensional ferromagnet. *Physical Review A 60*:252.

Kürten, K. E. 1988. Critical phenomena in model neural network. *Physical Letters A (Amsterdam) 129*:157.

Lapedes, A. and Farber, R. 1986. A self-optimizing, nonsymmetrical neural net for content addressable memory and pattern recognition. *Physica 22D*:247-59.

Lippmann, R. P. 1987. An introduction to computing with neural nets. *IEEE ASSP Mag. 4*:4-22.

McCulloch, W. S. and Pitts, W. 1943. A logical calculus of ideas immanent in nervous activity. *Mathematical Biophysics 5*:115-133.

Marcus, C. M., Waugh, F. R. and Westervelt, R. M. 1990. Associative memory in an analog iterated-map neural network. *Physical Review A 41* (6):3355-64.

Marcus, C. M. and Westervelt, R. M. 1989. Stability of analog neural networks with delay. *Physical Review A 39*(1):347-59.

Matsuba, I. 1989. Optimal simulated annealing method and its application to combinatorial problems. *Proceedings of the International Joint Conference on Neural Nets. 1*:541-5 IEEE #89CH2765-6.

Metropolis, N. et al. 1953. Equation of state calculations by fast computing machines. *Journal of Chemical Physics 21*(6)1087-1092.

Minsky, M. and Papert, S. 1969. *Perceptrons: An Introduction to Computational Geometry*. Cambridge, Massachussetts:MIT Press

Moussouris J. 1974. Gibbs and Markov random systems with constraints. *Journal of Statistical Physics 10*(1):11-33.

Mueller, P., van der Spiegel, J., Blackman, D., Chiu, T., Clare, T., Dao, J. Donham, C., Hsieh, T. and Loinaz, M. 1989. A general purpose analog neural computer. Proceedings of the International Joint Conference on Neural Networks. IJCNN '89 Washington, D.C. 2:177-82

Myerson, R. B. 1978. Refinements of the Nash equilibrium concept. *International Journal of Game Theory 7*(2):73-80.

Nash, J. 1951. Non-cooperative games. *Annals of Mathematics 52*(2): 286-95.

Nemoto, K. 1988. Metastable states of the SK spin glass model. *Journal of Physics* A: Math. Gen. **21**:L287-L294.

Ogden, C. K. and Richards, I. A. 1923. *The Meaning of Meaning*. New York:Harvest/Harcourt, Brace & World, Inc.

Pineda, F. J. 1987. Generalization of back-propagation to recurrent networks. *Physical Review Letters 59*:2229-32.

Rice, P. M. 1979. Local strategies and equilibrium. *International Journal of Game Theory 8*(1):1-25.

Rochester, N., Holland, J. H., Haibt, L. H. and Duda, W. L. 1956
Tests on a cell assembly theory of the action of the brain. I.R.E. Transactions on Information Theory. vIT-2 pp80-93.

Rosenblatt, F. 1960. Perceptual generalization over transformation groups. In *Self-organizing systems.* Edited by M.Yovits & S. Cameron. New York:Pergamon

Rosenblatt F. 1961. *Principles of Neurodynamics: Perceptrons and the Theory of Brain Mechanisms.* Cornell Aeronautical Labs.

_____________ 1962. *Principles of Neurodynamics: Perceptrons and the theory of Brain Mechanisms.* Washington, D.C.:Spartan Books.

Rumelhart, D., McClelland, J. and the PDP Research Group 1987. *Parallel Distributed Processing: Explorations in the Microstructure of Cognition*, Vols. 1 and 2 Cambridge, Massachussetts: MIT Press.

Sejnowski, T. J., P. K. Kienker and G. E. Hinton 1986. Learning symmetry groups with hidden units: beyond the perceptron. *Physica 22D*: 260-75.

Shaw, G. L. and K. J. Roney. 1979. Analytic solution of a neural network theory based on an Ising spin system analogy. *Physics Letters 74A*(1,2):146-50.

Shepherd, G. M. 1974. *The Synaptic Organization of the Brain*. Oxford, England: Oxford University Press.

Sompolinsky, H., Crisanti, A. and Sommers, H. J. 1988. Chaos in random neural networks. *Physical Review Letters 61*:259.

Spitzer, F. 1971. Markov random fields and Gibbs ensembles. *American Mathematical Monthly 78*:142-154.

Szu, H. and Hartley, R. 1987. Fast simulated annealing. *Physics Letters A 122 (3,4)*:157-62

van Enter, A. C. D. and van Hemmen, J. L. 1984 Statistical-mechanical formalism for spin-glasses. *Physical Review A 29*:355.

van Hemmen, J. L., D. Grensing, A. Huber, and R. Kuhn 1988. Nonlinear neural networks I: general theory. *Journal of Statistical Physics 50*(1):231-257.

von Neumann, J. 1961-63. *Collected Works* 5. Edited by A. H. Taub New York: Pergamon Press.

Waugh, F. R., Marcus, C. M. and Westervelt, R. M. 1990. Fixed point attractors in analog neural computation. *Physical Review A 64* (16):1986-89.

Wilson, G. and Pawley, G. 1988. On the stability of the traveling salesman problem algorithm of Hopfield and Tank. *Biological Cybernetics 58*:63-70.

Wolfram, S. 1984. Universality and complexity in cellular automata. *Physica 10D*:1-35.

Index